Previous Publications

Maartje Abbenhuis, *The Art of Staying Neutral: The Netherlands in the First World War* (2006)

Sara Buttsworth, *Body Count: Gender and Soldier Identity in Australia and the United States* (2007)

Restaging War in the
Western World

Restaging War in the Western World

Noncombatant Experiences, 1890–Today

Edited by Maartje Abbenhuis and Sara Buttsworth

RESTAGING WAR IN THE WESTERN WORLD
Copyright © Maartje Abbenhuis and Sara Buttsworth, 2009.

First published in 2009 by PALGRAVE MACMILLAN® in the United
States – a division of St. Martin's Press LLC, 175 Fifth Avenue, New York,
NY 10010.

Where this book is distributed in the UK, Europe and the rest of the
world, this is by Palgrave Macmillan, a division of Macmillan Publishers
Limited, registered in England, company number 785998, of Houndmills,
Basingstoke, Hampshire RG21 6XS.

Palgrave Macmillan is the global academic imprint of the above companies
and has companies and representatives throughout the world.

Palgrave® and Macmillan® are registered trademarks in the United States, the
United Kingdom, Europe and other countries.

ISBN-13: 978-0-230-61266-2
ISBN-10: 0-230-61266-0

Library of Congress Cataloging-in-Publication Data

Restaging war in the western world: noncombatant experiences,
1890–today / edited by Maartje Abbenhuis and Sara Buttsworth.
    p. cm.
Includes bibliographical references and index.
ISBN 0-230-61266-0
1. Military history, Modern—19th century. 2. Military history,
Modern—20th century. 3. Warfare. 4. War and literature.
I. Abbenhuis, Maartje M. II. Buttsworth, Sara.

D396.R47 2009
355.02—dc22   2008038409

A catalogue record of the book is available from the British Library.

Design by Macmillan Publishing Solutions

First edition: March 2009

10 9 8 7 6 5 4 3 2 1

Printed in the United States of America.

To our wonderful children, Joseph, Elijah, and Helena: may they never need to grow up to be soldiers.

And, in loving memory of Tobias Gabriel Ivimey, "Toby," 19 July 2008–10 August 2008. Sweet dreams little man.

Contents

List of Figures

INTRODUCTION

Onto Center Stage: Warfare in the Western World

Sara Buttsworth and Maartje Abbenhuis

We both have the joys and frustrations that are two-year-old boys—Elijah and Joseph. They are great friends and love to dance. Joseph is cool and likes jazz. Elijah's current favorite song, because daddy showed him the film clip, is Christina Aguilera's "Candy Man" from her 2007 *Back to Basics* album. Just as disturbing as seeing him mimic the dance moves is the recognition that, even through something as innocuous as a frivolous pop song, he is being exposed to representations of war, in this case the remediation of the Andrews Sisters' "Boogie Woogie Bugle Boy of Company B," which was a worldwide hit upon its release in 1941. Aguilera's version is a romanticization of the Second World War, not only through the reworked music but also through the interspersion of film clips depicting soldiers in Second World War uniforms with Aguilera gyrating in various costumes from the era, including one that replicates Rosie the Riveter, overalls, kerchief, and all. The sexism and sexualization of women in uniform and women in the workplace that were very much present in representations of women attempting to "do their bit" in the 1940s are also very much present in this representation that was created more than 60 years after the original song was released. There are no toy guns in our homes, or tanks, or warplanes, for that matter. We refuse to buy clothing that has camouflage print on it. But war has crept into the lives of our two-year-olds even if they are too young yet to recognize it.

The glamour and bright colors that attract Elijah to Aguilera's "Candy Man" are present in many depictions of war, as war is theater: pageantry, uniforms, drama, villains, victims, and heroes. Even if the colors used are actually the more drab hues of khaki and navy blue, participation in war, particularly if it can be written as a "just" war, is lauded in places as far apart as New Zealand and the United States. Cadets march on "parade" grounds while returned soldiers parade through cities to celebrate victories or commemorate wars long past. It has become a cliché for soldiers to describe war itself as 90 percent boredom and 10 percent terror. Perhaps unsurprisingly, soldiers' experiences and their activities in armed combat sit firmly at the center of how many westerners think about war. In many popular representations of war, soldiers are, more often than not, cast as the main actors. They are the heroes we aspire to be like, revere for their achievements and bravery, and commemorate for their sacrifice. Obviously, war is also about victims, those upon whom "war is inflicted" or to whom "war happens." They are often cast as hapless and helpless women, children, and the elderly fleeing the violence and suffering of the warfronts. If not seen as a faceless mass, noncombatants are frequently presented as the supporting actors, individuals to be pitied, cared for, and commemorated as "victims" and consequently depicted as having very little agency of their own.

But what is war for those behind the scenes, or in the stalls, for the understudies, the costumers, the orchestra, the families of cast members, the protesters of controversial productions, the reviewers, or even the support crew behind the counter at the bar? In an age where war has become increasingly more total for victims than perpetrators and evermore pervasive in daily life, many more noncombatants than soldiers feel the impact of war. The omnipresence of war in everyday life renders it almost pedestrian in Western culture, which is partly due to the removal of distance between the ready availability of images and information about wars, both real and imagined. As Margaret Higgonet so beautifully expressed it in her discussion of war toys and narratives of the Great War:

> War toys remind us that war is not something that happens far away on a neatly contained "battlefront" but part of the everyday; war is not simply an isolable event but an eruption within a continuum that runs through what we call peacetime.[1]

This is as true for the tin soldiers that were the toys of the early twentieth century as it is for the memorabilia on sale at various fairs and conventions in the early twenty-first century. The weaponry, uniforms, and mementoes of wars past are the conduits for the continued pervasiveness of war in the present and lay the basis for the continued primacy of war in our culture into the future.

Our book cover illustrates the integral place war themes have in everyday life so well. The photograph, originally one of a triptych on the same scene, was taken in 1907, in Ireland, of brother and sister, Rex and Gillian, posing in their best clothes. Rex is at play with his miniature infantry, cavalry, and artillery replicas carefully lined up and arranged, while Gillian sits slightly apart from her sibling, looking on. She is physically removed from the staging and reenactment of war, but she remains a suitably unimpressed witness of her sibling's war play. In what is obviously a posed photograph, the use of the military toys (rather than other nonmilitary ones) and the setup of the two children in a traditionally gendered way, with the sister as an audience to but not a participant in her brother's war games, indicates how integral an imagined idea of war was (and remains) to the Western world's way of life and ideas about themselves and their society. While we seek to protect the "innocence" of our children and shield them from ever experiencing the "reality" of war, staged and unstaged wars are as much a part of their lives as they were for these besmocked children photographed 100 years ago.

Although, throughout the twentieth century, war has often been viewed more as hellish horror than romance, this too is a part of the theatricality of combatant war stories that have overshadowed alternative scripts. There is a strong emphasis on the British soldier poets of the First World War in the English literature curricula of Great Britain, Australia, and New Zealand, providing an almost singular perspective on war and soldiering to audiences distanced from the mud and trenches of that conflict by a century. The marches and memorials to the First World War leave little room for public discussion of the treatment of conscientious objectors or the legacies of imperialism. For example, the transition of the "real" soldier Audie Murphy to film star, best known for playing himself in the film based on his Second World War memoir *To Hell and Back,* is a fascinating example of the transition from combatant to celluloid hero and of the public's fascination with war stories as dominant narratives of the twentieth century.[2] The possibility of, and capacity to, accurately depict the carnage of war in cinema remains an obsession present right up to today, where this carnage is mingled with the idealized romance of retrieving lost men as in Steven Spielberg's *Saving Private Ryan.*[3] Popular culture is dominated by images of war, the bonds between soldiers, and the spectacles of battle. From the thunderous volley of cannons in Tchaikovsky's *1812 Overture* to the melancholy and horror of Wilfred Owen's poetry, from the constant diet of films about war and soldiers to the fascination with uniforms that not only permeates adult fashion but has seen the production of pink camouflage baby rompers. Even the most apparently peaceable pursuits have been pervaded by symbols, sound bites, myths, and stories about war. War as an object of fascination with military matters or needless suffering often ignores the views and voices from the wings: the voices of those who witness or experience war as noncombatants or nonparticipants. Furthermore, the fascination with war as a legitimate pursuit can obscure the ways in which it is normalized into everyday life. Yet it is precisely these voices and influences that are myriad and crucial to a broader understanding of the central position that warfare has held in Western culture since the late nineteenth century. One of the main aims of this collection is, therefore, to investigate the ways in which war and militarism remain central foci points in Western culture. In so doing, we highlight the ways in which the assignment of primacy to soldiers in the analysis and representation of warfare undermines the recognition of the ways wars affect noncombatants both during periods of conflict and in their aftermath. We seek particularly to emphasize that war is an all-consuming part of modern Western culture and should be analyzed as a multifaceted phenomenon that is as important to noncombatants as it is (or was) to combatants. This is not, of course, to suggest that there is no scholarship available on the impact of war on civilians or that war studies have been one-dimensional up to this point in time. Quite to the contrary: there is a growing and massive literature available on warfare and culture, war and memory, and war and society studied from numerous perspectives and approaches. We hope that this collection adds to this scholarship by reiterating the pervasiveness of war themes in everyday life and asserting the prominence of the impact and role played by wars in the lives of noncombatants in the modern world. Even more broadly, our ambition is to reposition the traditional place assigned to civilians and noncombatants, namely, from the peripheries of actual and imagined "theaters of war" to the center stage. In the process, we hope to reinvestigate the role played by martial prowess in Western culture; the concept of victimhood in postwar memory and culture; the long-term legacies of wartime events as they influenced noncombatants; the conceptualization of war, militarism, and conflict in cultural representations for children and adults; and the relevance of peace activism in war and peacetime.

Overview

> The simmering, inchoate conflicts that characterize the contemporary world, conflicts no longer on focused territorial disputes between nations but on claims for cultural and political recognition and the distribution of resources, make it ever more difficult to imagine an impermeable private sphere, safe from the psychic impact of global dislocation and violent disruption.
>
> Clair Wills[4]

This collection seeks to move the perspectives that are marginal to mainstream war stories and histories onto center stage, acknowledging their concept of "total war" having a longer-term impact is rarely pursued. The influence of total war in peacetime, in the aftermath of conflict, deserves much more attention. Both world wars (and not only these two conflicts) had numerous everyday and long-term legacies. It took people, some of whom had no direct experience of or involvement in the war, out of ordinary situations into the extraordinary. For others, it made the extraordinary commonplace. Gabrielle Fortune's chapter in this collection on war brides makes explicit these connections between the ordinary and the extraordinary and the ways in which war became a defining identifier for some women throughout the rest of their lives—in spite of some of them never having witnessed warfare directly. Similarly, Ismee Tames focuses on the problems of constructing a new identity for the children of Dutch Nazi collaborators in the immediate aftermath of the Second World War. Their status both as children and as children of a neutral nation should have protected them from war and its impacts. The realities of the Second World War's violation of both Dutch neutrality and the "sanctity" of childhood through their parents' politics, however, made it impossible for many of these children to begin or continue new lives, untainted by their wartime experiences, once the war was over. For many, the outbreak of peace does not mean the end of the war.

In many respects, the intensity with which war themes continue in peacetime suggests that the scholarly obsession with periods of "total war" may need to be broadened. We see a need to understand the existence of real and imagined wars as a historical constant, as a theme that has influenced the Western world for at least the last 200 years, if not longer, and to take the analysis of war beyond the obvious links between military conflict and soldiers and between noncombatants and victimhood. The concept of "total war" itself has, it seems, tied scholarly attention to extraordinary periods of participation by noncombatants in warfare, such as occurred in the two world wars. Admittedly, this has enabled scholars to understand war as more than a military phenomenon, a very important development and one that is supported by several of the contributions in this volume. However, the label "total war" also suggests that there is something extraordinary about the seconding of society to war, that, in fact, war is meant to be for soldiers, not for noncombatants.[10] In a rather perverse way, the fascination with periods of perceived "total war" and war crisis, such as the First and the Second World War, has diverted attention away from the continuities of war in supposed "ordinary" "peaceful" times.

It is difficult to define the concept of "peace". In its most simplistic form, "peace" can best be classified as "the absence of war". We contend, however, that in the Western world in the modern era, an absence of war does not

Figure 0.1 Charles Upham, the New Zealand Second World War veteran and double–Victoria Cross awardee, used to advertise photocopiers on this giant billboard visible in central Auckland, New Zealand, November 2007.
Source: Photograph courtesy of Gabrielle Fortune, Private Collection

exist. War is omnipresent and an everyday reality. Even in peacetime, many Western communities are obsessed by war themes. This is as true for the United States in the late nineteenth century as it was for participants in the antiwar movements in Scandinavia in the interwar years, as it was for many countries during the Cold War era, and still is in the post-9/11 world. So, while many Western nations have lived in relative peace for many decades and only indirectly or haphazardly experience warfare—one would not claim, for example, that the United States today, while it is officially at war in Iraq and Afghanistan, is a total war society—the fascination with all things war and warrior remains.

Recent examples that we have come across of this warrior obsession include an advertising billboard in Auckland, New Zealand, for photocopiers (see figure 0.1). This giant canvas prominently situated in the middle of Auckland's city center aligns the experiences of the New Zealander and double–Victoria Cross awardee Charles Upham with the concept of heroism and leadership. The advertisers are promoting their photocopiers as leaders (not followers) by plugging into what they expect mainstream New Zealand understands of Upham's Second World War experiences. The billboard is as much about national pride and achievement as it is a representation of the deep-seated fascination with military heroism in everyday New Zealand culture. Furthermore, the advertisers are indirectly and unintentionally profiting from the (unrelated) theft of one of Upham's medals (along with those of other prominent New Zealand veterans) from the Waiuru Army Museum in December 2007.[11] The theft received media saturation in New Zealand, particularly as an unprecedented NZ$300,000 reward was offered for information leading to the recovery of the medals—put up by private donations and supported by the New Zealand police.[12] The theft outraged the country. Most New Zealanders felt that the thieves had committed an unpardonable act, an affront to the nation, to the families of the medal recipients, and to the memory of the sacrifice made by these brave soldiers. The connections between the two world wars, their memorabilia, and concepts of New Zealand and national identity are still very much alive and well.

An even more troubling example of the place military heroism plays in modern society is the annual "War and Peace" military collectables show held at Beltring in Kent, United Kingdom.[13] There is very little focus on peace at this massive event, which sees military reenactors, war paraphernalia collectors, and the British armed forces combine their talents to present a host of ways in which to promote and sell all manner of things to do with the military and wars, past and present. It would actually do to rename the show "War *in* Peace." The *Scale Model Collector* magazine raved about the 2007 event in the following terms:

> the show welcomed over 5000 military vehicles. . . . Waffen SS, Tommies and GIs from WW2 to the present day were dug in around the encampment; each living life as close to their theme as possible. There was even a Vietcong tunnel complex leading into the Vietnam experience, through which visitors could crawl. The centre-piece of the show was the main arena, where infantry and armour had the chance to show off their talents bringing history to life. . . . You can walk into a stall and come out as Waffen SS "Landser" or a "Band of Brothers" US paratrooper. . . . Beltring is one of those events that the whole family can enjoy. Dad dressed in uniform, while mum and the kids walk around the event in similar period clothing with children suitably labeled and carrying their boxed gas-masks. . . . [T]here was plenty of time to party in the "Victory Marquee" where visitors were adorned in floral Hawaiian shirts, and sailors sporting US Navy hats made this family friendly place a cross between "Butlins" and "Pearl Harbor" with live 40's music and competitions.[14]

Tellingly, the photographs that accompanied the article included laudatory captions like "The SAS fast attack vehicle looked fun to drive" and "The

Germans [in the Second World War] had nice uniforms." This event, now in its twenty-sixth year, more than anything else illustrates how "bloodless warfare" has caught the imagination of sections of modern society. The organizers, participants, and visitors to Beltring seem to be there for the "idea" and "romance" of war, rather than for attaining any sense of its reality.

In the last twenty years, under the influences of feminist scholarship, cultural studies, Holocaust studies, and social history, histories of war have developed, and there have been many far-reaching monographs and collections that explore the relationships between media and war, gender and war, childhood and war, the impact of peace movements, and the importance of neutrality.[15] To this end, some of this collection revolves around the concept that war has become a "spectator sport,"[16] a bloodless affair, like the Beltring extravaganza discussed above, with a potential audience of millions. But even the "unreal" experience of mediated war can have important and long-lasting consequences. Some of the contributions in this collection focus on individuals, groups, and communities that have worried and philosophized about and challenged the possibility of war occurring in their society—people who obsess (or have obsessed) about the need for peace and the difficulties of removing the existence of war. In this sense, it is about those who *think* about war and the possibility of future conflict and act to mitigate it. Thinking about war is a complex, yet common, sociological phenomenon. Above all, however, this collection looks at the layered and long-lasting impact of actual and imagined wars on individuals and groups whom we would not traditionally consider as being agents or targets of warfare but whose lives, nonetheless, were (or are) fundamentally altered by the existence of such conflicts. What is important and unique about this collection is the way in which we seek to draw together this multiplicity of approaches across disciplines, historical periods, geography, theoretical directions, and subject matter. Each chapter deals with voices that have not been heard and yet for whom war was (or is) a defining experience. All of the articles present perspectives normally excluded from narratives about war. They also add to the growing body of work seeking to broaden the discursive boundaries and understanding of a phenomenon that penetrates the lives of noncombatants in an age where it is increasingly problematic to be "at peace."

An outcome that was not intentional was the volume of content we have on the impact of war on children whether immediate or representational. The chapters by Ismee Tames, Sara Buttsworth, David Rosen, and Karen Hall all deal with the ways in which children and ideas about childhood do not remain sacrosanct in the face of the centrality of war. Another recurrent theme throughout the book is that of activism against war or by or on behalf of the "victims" of war. Penelope Adams Moon and Suellen Murray both explore the difficulties and challenges of women protesting for peace in two very different Cold War contexts. Hall's chapter introduces the idea of organizing peace activism through play. Irene Andersson unravels the complicated position committed pacifists have when their own safety is threatened, and Rosen introduces the difficulties of dealing with forms of advocacy or activism that do not recognize the agency of children in times of war.

In terms of subject matter, this book is not meant to be an end point. We hope it inspires and propels further work in the field and diversifies scholarly discussions about the place and context of warfare in the Western world. In a work of this kind, so broad in scope and subject matter, there are many areas we could have covered and numerous obvious omissions. For example, we have not included a chapter on the designation of servicewomen as noncombatants in many modern armed forces. This omission was not by design, although we have to some extent attempted to engage more exclusively with voices that do not come directly from formal military institutions. Likewise, it would have been appropriate to include a chapter on recent wars (particularly post-9/11) or on contemporary media in presenting and utilizing war themes, including blogging, Internet reporting, and war gaming. These are all themes that deserve more attention in the scholarship on war and peace. What we have tried to do is present an array of interesting and divergent views on warfare from a variety of historical periods and geographic locations. The variation in the collection makes our point about the centrality of warfare in Western society all the more poignantly.

It is important to note that beyond the specification of "'western," this collection is not really about geography. We have included diverse studies from all over the Western world—ranging from Europe to North America and the Pacific. While acknowledging that there are obvious contextual differences, it is the seemingly eclectic nature of the contributions that illustrates the ubiquity of war in the Western world, regardless of physical distance between an actual "theater of war" and the people who are affected by it. Obviously, we have not been able to include chapters on every country considered to be "Western." Still, many of the chapters are about countries that have experiences with the same conflicts or have exposure to similar media content. Perhaps, most conspicuously, we have no direct contributions about Britain, France, or Germany. This was unintentionally done but can be seen as a reflection that the "Western" world extends well beyond northwest Europe. It also does not mean that these important countries are absent from the discussions offered here. For example, many of the war brides discussed by Gabrielle Fortune were British and carried their nationality proudly to their new homes in New Zealand. The permeability of European borders is also amply demonstrated by Ismee Tames in her discussions of the travels of Dutch children in and out of Germany during the Second World War. While particular nations are the specific focus of many of the chapters, the frameworks and interests of our authors demonstrate the impossibility of keeping discussions of war within national or temporal boundaries.

Chapter Outline

The first chapter in the collection, by Mark Potter, focuses on the attractions of warfare in peacetime, namely, in post–American Civil War New York society. It highlights many of the themes that run through the subsequent chapters about martial prowess and the romanticism of soldiering that is still so prevalent in Western society today. Potter shows that the grief and mourning that existed in the years immediately following the end of the Civil War in 1865 also witnessed the romanticization of one of the bloodiest and most mechanized conflicts the world had yet seen. The legacy of war inspired young men, supported by their families, to join militias. It was not impending war, but a war that was over, that stirred such displays of peacetime military aptitude. In the years that followed, the memorialization of war became a family affair as well as a spectator sport, in spite of (or perhaps because of) the grief that continued unabated for many. Public mourning and the spectacles that accompanied it continued long after these young men put aside their drilling uniforms. The Civil War and its legacies continue to live on in the twenty-first century through battlefield tourism, traditions of which possibly had their roots in the would-be militiamen of this chapter.[17] Today, battlefield tours, not only of American Civil War sites, but also those of other major conflicts, particularly the First World War, are now a lucrative and key part of the holiday industry's increasingly popular "grief tourism" profile.[18]

Where Potter focuses on the idealization of war as a worthy enterprise, Irene Andersson, in her chapter on the peace initiatives undertaken by Social Democratic women in Sweden, looks at ways in which the idealization of peace (and the avoidance of war) underpinned the activities of a large group of women at the outbreak of the Second World War. Andersson tells a story not often heard outside Sweden. She examines the political and personal turmoil caused when a neutral country is threatened by a bellicose and determined belligerent. While historians traditionally depict neutral countries in the 1930s as hiding from the international arena and misunderstanding the realities of pending European conflict,[19] Andersson's chapter amply demonstrates the very real understanding and fear of war's potential impact by citizens of one such neutral nation. The Social

Democratic women, she discusses, saw themselves as "guardians of neutrality," a concept that they imbued with distinctly moral and pacifist ideals. Sweden's neutrality was not only a practical foreign policy option, with the aim of protecting the country from invasion, it was also part of the Swedish national character, a sacred dogma that focused on neutrality as a morally upstanding way to conduct international affairs. This ideology allowed Swedes to think of themselves as operating "above" and "beyond" war. Of course, in reality, neutrality could not protect a nation or a society from war, and Andersson carefully illustrates the impact that the prospect of being forced to go to war had on these women, who were ideologically wedded to peace activism and neutral pacifism. She shows how conflicted many Social Democratic women were when necessity dictated they must support civil defense measures and other war preparations. Andersson's research is a rich resource for anyone interested in pacifism, gender, and war: these women saw their pacifism as a political choice tied to their identity *as women,* and, even more than that, as *Swedish* women.

By 1939, Swedes were certainly not blind to the realities of war or ignorant of the possibility that they may fall victim to the conflict that was escalating around them. The idea, then, that Sweden was a naïve player in the international arena is as problematic as the assumption that neutrality precluded sophisticated debates about the role of the military and war in society. As Andersson shows, Social Democratic women in their publication *Morgonbris* were engaged in multifaceted discussions about the place of war in the world at large and in their neutral nation particularly. Obviously, there was more possibility to openly protest against war in a neutral country, but neutrality and pacifism did not necessarily go hand in hand. In fact, in Sweden, the Federation of Women Social Democrats was deemed radical for not doing more to promote defense duties among its members. Sweden was not exceptional in this respect as in both the First and the Second World Wars the European neutrals had to mobilize their armed forces and implement all manner of defense mechanisms to protect their nonbelligerency. Even though these societies did not actually fight, the demands of civilian society were often subsumed under the pressures of military mobilization. As a result, it was as impossible for Sweden's politically minded women to avoid issues of war and militarism as it was for women in belligerent societies. Andersson's chapter provides an important addition to the meager literature on neutral societies in wartime and the, even more meager, scholarship on neutral societies in peacetime.[20]

Gabrielle Fortune's contribution, much like Andersson's, highlights how private individuals were forced to compromise their ideals and their lifestyles due to the presence of war, even when they themselves were not active participants in the conflict. Fortune's examination of war brides who came to New Zealand in the aftermath of the Second World War focuses on the difficulties many of these women faced leaving the familiarity of home to make new lives with new families in a new culture halfway across the world. Fortune's chapter indirectly highlights the role warfare plays in breaking down distance and how cultural barriers were altered and molded by the war experiences of a wide variety of individuals. While New Zealand has always been a nation of immigrants, the postwar impact of foreign-born war brides on New Zealand society was significant. Far more tellingly, the war had a fundamental impact on the women themselves. The stories interwoven throughout Fortune's chapter reveal the personal aspects of the impact of the war and the feelings of displacement that have followed many of the women ever since. The Second World War may have ended in 1945, but it remains a key part of these women's identities right up to the present day. While soldiers may become "old soldiers" or former soldiers, the war part of these women's identities lingered long after they ceased being new brides—"once a war bride always a war bride." Firsthand experience of combat was not a requirement for the war to make an ongoing impact on these women's lives.

The same conclusion—about the ongoing and fundamental impact of war on individuals' lives—can be drawn about the Dutch children of Nazi collaborators that are the subject of Ismee Tames's chapter. Tames' new research has ramifications for the way we perceive and discuss the long-term consequences of war on identity and the legacies of victimhood. Her chapter shows us how our understanding of children—as innocents to whom no "bad" things should happen, particularly not war—underpins our perceptions of the impact of war on children. Tames' children of Nazi collaborators were not all innocents during the Second World War nor can they be cast solely as victims. However, in the postwar era, their experiences have been co-opted into a larger discourse on the legacies of "the war" in the Netherlands itself.

Until quite recently, the ways in which historians and the wider public conceived of their nation's period of Nazi occupation was in terms of *goed* (good) and *fout* (literally, "wrong," by implication "on the (morally) wrong side").[21] In this conceptualization, everyone who fought or resisted the Nazis or who was victimized by them was *goed*. Those who, in any way, supported the occupiers was *fout*. In the immediate aftermath of the war, *foute* Netherlanders were identified and labeled. This identity remained (and, in some cases, still remains) with them. Their children also suffered and came to be associated, labeled, and identified with the "wrongs" of their parents (regardless of their own very real and actual war experiences). As adults themselves, some of them published memoirs about their war experiences and postwar lives. Tames shows that in doing so they criticize wider society for victimizing and punishing them for the wartime "sins of their parents." What Tames' contribution asks us to do—as do those of Rosen and Buttsworth in subsequent chapters—is to reinvestigate the place of children in war and the legacies of war on our memories of childhood. Tames also shows how collective memories and national myths of particular conflicts can drown out and silence the voices of divergent war experiences, a theme that is echoed in Buttsworth's chapter.

While on the face of it David Rosen's analysis of literary representations of child soldiers focuses on the idea of children as warriors (or combatants), Rosen's point is that even in literature about child soldiers the (Western and modern) idea that children cannot (and should not) be warriors remains. Rosen shows that Western discourses of advocacy begin with the premise that the construction of childhood innocence should be taken as a given and that children are always the victims of war. Such discourses, while absent from French and American literary representations of the eighteenth and nineteenth centuries right up until the First World War, are very much present in representations of child soldiers in Africa in the late twentieth century.[22] In an interesting parallel with Potter's work, Rosen discusses the figures of heroic child soldiers in fiction about the American Revolution that was written during the Second World War. The character "Johnny Tremain" seems very much a part of the tradition that saw adolescents and young men don uniforms after the Civil War and represent the nation's future military prowess. Rosen illustrates that the discourses dictating that sanctions should be directed against adults and not children are mired in the cultural specificities that define exactly what is a "child." Here he helps to contextualize the notion of "innocent childhood" that Tames uses as the integral theme to her chapter.

Rosen has skillfully woven connections between the real problems dogging the United Nations' attempts to protect children who become soldiers and their representation through literature. To do this he has examined three novels, all published for the Western market: *Beasts of No Nations* by Uzodinma Iweala, *Moses, Citizen and Me* by Delia Jarrett-Macauley, and *Johnny Mad Dog* by Emmanuel Dongala. In these works, the heroism of the child soldier from earlier literature is absent and, instead, we find traumatized children whose psyches are even more wounded than their bodies. The agency of these children is always in doubt in these works, a reflection of the ways in which recent histories have also dealt with children and war as Rosen pointed out in his recent book *Armies of the Young*. The ideal child is never a combatant, but we do not live in an ideal world, and combat is very real for many children. Are these children then corrupted, no longer "ideal," and, therefore, no longer worthy of protection? Or is it more important to acknowledge that the impact of war is not containable for children or adults? Where Rosen uses children as subjects in adult literature about war, Sara Buttsworth, in her chapter on the picture book *Lottie: Gallipoli Nurse*, assesses the ways in which war is represented in children's literature in New Zealand in the early twenty-first century. In her examination of *Lottie*, Buttsworth highlights how wartime mythologies underpin the way in which New Zealand children are educated about their national history and the relevance of war in it. She explains how ANZAC[23] mythology, which centers on the male (and largely white) combat soldier, his bravery, and fighting prowess in the face of the insurmountable odds at Gallipoli in 1915, remains central to New Zealand's national identity (which is, incidentally, also true in Australia) in spite of its exclusionary and imperial origins. Her chapter analyzes how a picture book ideally placed to question and subvert the ANZAC myth actually reinforces it and, in doing so, offers its youthful audience no historically valid ways of "reading" about war or the past. Despite initial appearances, *Lottie* is not a story about professional nurses during the First World War. Instead, Lottie is a prop in her own story, which is a retelling of the ANZAC myth. Even more worrying is the questionable use of archival material by the author of *Lottie* and the anachronistic and frequently careless illustrations that accompany the written text. Buttsworth admonishes the author, illustrator, and publisher for allowing such carelessness and inaccuracy to creep into this work of "faction," when it is precisely children who need to obtain a balanced, accurate, and nuanced view of war. Buttsworth presages the strong activist message in Hall's chapter on war toys, namely, that by not offering children a complex view of war and its role in society, we are in danger of seeing our children replicate and support the grand narratives and legends of war on which the many justifications for war in the Western world are built. What hope can children have of avoiding war in the future if they have no understanding of the multiple contexts of war in the past?

The last three chapters examine the experiences of pacifism and "peacetime" activism, themes that Andersson also addresses in her chapter on Sweden in the interwar period. Both Penelope Adams Moon and Suellen Murray look at women's activism for peace in the latter part of the twentieth century. The alignment of women and peace movements in this collection was not premeditated, and we by no means wish to extend the binary stereotype that the Swedish Social Democratic women were so fond of— that women are essentially more predisposed to pacifism than men. However, the connections between certain forms of peace activism and second-wave feminism in the twentieth century are worth teasing out as further examples of the ways in which war impacts upon certain groups that have often been defined as noncombatant by virtue of their gender.

Beginning with a shocking and poignant story of a woman who set herself on fire to protest the war that was beginning in Vietnam, Penelope Adams Moon's chapter "We Aren't Playing that Passive Role Any Longer" examines the struggles of women peace activists during the Vietnam War. In a protest culture that focused almost exclusively on the draft, women who were committed to peace found themselves ignored or working on the fringes. The combat soldier, or the potential combat soldier, marginalized all other voices and issues. As Adams Moon points out, women who protested against the Vietnam War did not always do so in different ways to men, but they did choose methods and activities that had relevance to their lives and to their understanding of being women. Crucially, Adams Moon establishes different kinds of feminine identity depending on class, race, ethnicity, and religion. All these women were opposed to the war, but they had slightly different reasons and different tactics depending on who, and where, they were. This research is extremely important in establishing the difficulties women have participating in public politics in general and the politics of wartime in particular. The gender biases that privileged the male combatant soldier also privileged the male peace activist, lending many protests a militant air in spite of their pacifist objectives. Women worked hard not only for peace but also for recognition that they too were affected by the war and that they too had a right not only to speak but also to be heard. Adams Moon's work indicates too that there were as many divergent ways to experience peace and peace activism as there are ways to experience warfare.

Twenty years later in Australia, Cold War peace activism took on different forms again when other groups of women's peace activists faced similar struggles when they protested nuclear war and the presence of American nuclear naval ships in Australian waters. Suellen Murray, who was one of these activists, provides both an insider's perspective and an acute analysis of the tactics the media and politicians used to trivialize and caricature the activities of these women. Murray looks at two geographically separated camps that both protested the presence of nuclear weaponry on Australian soil: the Pine Gap Women's Peace Camp, held in central Australia in 1983, highlighted the presence of the United States base near Alice Springs, and, in 1984, the Sound Women's Peace Camp, held south of Perth that opposed the presence of U.S. naval ships with nuclear capacity. The peace activists in Australia were a part of a bigger movement that also encompassed the protests at Greenham Common in England in 1981. The Australian media tended to trivialize the activities of the women and portrayed them as either maternal and irrational or radical and irrational. These women, who cared not only about their own lives but also about their country and the planet, were portrayed as the lunatic fringe rather than as concerned citizens with a legitimate message. The politics of potential nuclear war were global in impact, and the women involved in antinuclear peace in Australia were acting locally against the impact of a war that would eliminate the need for soldiers altogether and indiscriminately wipe out entire populations.

"The War at Home" by Karen Hall brings together the themes of childhood and activism in her examination of the impact of military and war-inspired toys and games. Clearly polemical in intent and tone, Hall urges us, as parents, to embrace and promote ambivalent war-play by our children and, by doing so, demolish the myths of "heroes" and "villains" upon which Western ideas about war are built. In this way, we instead focus on the destructive potential of the military-industrial complex that underpins the reality of war today. Hall acknowledges that the incursion of war toys into the playroom is a way of desensitizing children to war and conditioning them to its "normalcy." Much like the spectacle of war that involved the whole family in America's "Gilded Age," playing at war has continued and is a part of many playrooms and computer consoles around the world. Future soldiers and citizens who accept the myths of "just wars" are still made through play and display nearly two centuries after the National Guard took part in its battlefield reenactments in the wake of the American Civil War. During the late twentieth and early twenty-first centuries, war has increasingly been commodified and is sold as a series of products to adults and children alike through the films we see, the television programs we watch, the games we play, and the toys we buy. What Hall wishes us to do, however, is to recognize that precisely because we cannot shield our children from the omnipresence of war in our media-saturated societies, we should teach them to appreciate more than the binaries of "Us" versus "Them" so engrained in our ideas about warfare and raise them to be equally skeptical of "just war" mythologies. The vehicle Hall utilizes to promote these issues is an analysis of the film *Small Soldiers* (1998) and its merchandising that representationally and literally brought warfare into the domestic sphere of the home.[24] *Small Soldiers* not only pokes fun at the role of war in society, it also made its audience, young and old, rethink their position on militarism. Hall uses the film as a way of encouraging different kinds of narratives and attitudes about war and war-play.

Conclusion

Children lose their youth too soon
Watching war made us immune.

The Dixie Chicks[25]

In 2003 in the wake of the U.S. invasion of Iraq, Natalie Maines, the lead singer of the Dixie Chicks, made an off-the-cuff statement to a concert audience in London, disparaging President George W. Bush and by implication this unpopular war. As a result, many country music stations in the United States took the Dixie Chicks off their playlists, people protested, and the band received death threats. Maines was told she had better "shut up and sing or my life would be over."[26] The "freedom of speech" that is supposedly one of the tenets of U.S. society was only "fine if you don't do it in public."[27] Other than making overt the kinds of censorship that were unofficially happening, the Dixie Chicks were also singing of a society that is groomed for war. The United States is not alone in this, as far away as New Zealand a century-old war also remains in public memory as the key definer of national identity. Similarly, in the Netherlands, a country proud of its historical neutrality, the Second World War is still the litmus test for "good" and "bad" Dutch citizens. War, even long after the guns are silenced in a particular conflict, continues to affect noncombatants and former combatants on individual, societal, national, and global levels in direct and more subtle mediated ways.

We have made explicit some of the connections between the different contributions to this book. We wish to stress, however, that we also sought out an eclectic array of contributions to reflect the eclectic and chaotic impacts that wars have, and have had, at myriad levels. While we, personally, have never been directly affected by war or exposed to combat, we are repeatedly affected by war and by the obsession our society seems to have for all things martial. Our children are also not exempt from the valorization of military culture and particular kinds of war stories. Will there be alternatives to *Lottie: Gallipoli Nurse* by the time Joseph and Elijah are in middle school? Will First World War myths continue to dominate peacetime twenty-first century New Zealand? In an article on wartime photojournalism that we would dearly have liked to have included in our collection, Wendy Kozol cites Andrea Liss with specific reference to the Holocaust: "[T]he demand to never forget is not directed at survivors, who can never forget, but at those who never experienced the events."[28] But what is it exactly we are being exhorted "never to forget"? That millions died in conditions most of us cannot even conceive? Certainly. But what about the complex social and political circumstances that led to these atrocities? It is only a very particular kind of story that future populations and populations far removed from actual events are asked never to forget. Still, in being asked, even if they are so very distant, these noncombatants are also affected by the continuities, if not the totalities, of war.

Notes

1. Margaret Higgonet, "War Toys: Breaking and Remaking in Great War Narratives," *The Lion and the Unicorn*. 31, 2007, p. 119.
2. Jesse Hibbs, dir., *To Hell and Back*. Universal International Pictures. 1955.
3. Steven Spielberg, dir., *Saving Private Ryan*. Amblin Entertainment. 1998; Albert Auster, "*Saving Private Ryan* and American Triumphalism," in Robert Eberwein, ed., *The War Film*. New Brunswick, 2005, pp. 205–213.
4. Clair Wills, "The Aesthetics of Irish Neutrality during World War II," *Boundary 2*. 31, 1, 2004, p. 57.
5. We have purposely used the word "his" in the book as one of the continuing assumptions about warfare is that, in spite of evidence of women's participation in historical wars and growing numbers of women in a larger range of specialties in western militaries, soldiering remains a masculine pursuit.
6. Jean Baudrillard, *The Gulf War Did Not Take Place*. Bloomington, IN, 1995.
7. Ken McLeish, "Playing War," *Interculture*. 5, 1, January 2008, p. 17.
8. A recent New Zealand example of this: Gary Sheffield, "Britain and the Empire at War 1914–1918: Reflections on a Forgotten Victory," in John Crawford, Ian McGibbon, eds., *New Zealand's Great War: New Zealand, the Allies and the First World War*. Auckland, 2006, pp. 30–68. Sheffield denounces recent changes in the First World War historiography that attempt to examine broader causes in favor of a focus on "King and Country" and the bravery of New Zealand soldiers. The constant media barrage from the Iraq War, which includes journalists "embedded" with certain U.S. military units, is an illustration of the endless fascination for certain kinds of footage and the active exclusion of other footage by military personnel and politicians. The ways in which certain stories have been created through particular kinds of "spin" seeks to glorify wars when public opinion is no longer as favorable, as can be seen in the saga of Pfc Jessica Lynch (Sara Buttsworth, "Who's Afraid of Jessica Lynch? Or, One Girl in All the World? Gender, Heroism and the Iraq War," *Australasian Journal of American Studies*. 24, 2, December 2005, pp. 42–62).
9. Hew Strachan, "Essay and Reflection: On Total War and Modern War," *International History Review*. 22, 2, 2000, pp. 341–370; John Horne, "Civilians and Wartime Violence. Towards an Historical Analysis," *International Social Science Journal*. 54, 4, 2002, pp. 483–490.
10. Of course, there is a vast literature available on the Second World War relating to civilians as victims of war, particularly on the topic of the Holocaust and its memory and commemoration in the postwar era. Importantly, Holocaust studies is one of the few fields that systematically acknowledge and analyse the long-term impact of war themes on noncombatants.
11. The advertisement predates the theft of the medals.
12. Oskar Alley, "Victoria Crosses stolen from museum," *The Dominion Post*. 2 December 2007, available at www.stuff.co.nz/4306566a10.html. Accessed 17 January 2008; "$300,000 reward for military medals return," *Dominion Post*.

17 January 2008, available at www.stuff.co.nz/4360672a6000.html. Accessed 17 January 2008.

13. The War and Peace Show website www.thewarandpeaceshow.com. Accessed January 2008. The following videos available on *YouTube:* www.youtube.com/watch?v=5GAG9Diun_Q&feature=related and www.youtube.com/watch?v=jpE PtKLDCFw&feature=related. Accessed January 2008.

14. Ade Pitman, "The War and Peace Show: A 'Belter' at 'Beltring; the World's Largest Military extravaganza,'" *Scale Models Collector International.* 37, 439, October 2007, pp. 72–73.

15. For examples by contributors to this collection: Maartje Abbenhuis, *The Art of Staying Neutral: The Netherlands in the First World War, 1914–1918.* Amsterdam, 2006; Sara Buttsworth, *Body Count: Gender and Soldier Identity in Australia and the United States.* Saarbrücken, Germany, 2007; David Rosen, *Armies of the Young: Child Soldiers in War and Terrorism.* New Brunswick, NJ, 2005 (which has generated heated discussion on H-War); Karen Hall, "Photos for Access: War Pornography and US Practices of Power," in Nico Carpenter, ed., *Culture, Trauma & Conflict: Cultural Studies Perspectives on Contemporary War.* Cambridge, 2007; "Consuming Witness: Combat Entertainment and the Training of Citizens," in Frances Guerin and Roger Hallas, eds., *The Image and the Witness.* London, 2007. Other recent collections in the area of cultures of war and peace include Jay Winter, *Remembering the War: The Great War between Memory and History in the Twentieth Century.* Michigan, 2006; Kathy Phillips, *Manipulating Masculinity: War and Gender in Modern British and American literature.* Basingstoke, UK, 2006; T. G. Ashplant, Graham Dawson, and Michael Roper, eds., *The Politics of War, Memory and Commemoration.* London, New York, 2000; Michael Neiburg, *Warfare and Society in Europe since 1898.* New York, 2004.

16. Colin McInnes, *Spectator-Sport War: The West and Contemporary Conflict.* London, 2002. To be fair, 9/11 has broken through the barrier of war as a "bloodless" affair somewhat.

17. Chris Ryan's recent book (*Battlefield Tourism: History, Place and Interpretation.* Oxford, Amsterdam, 2007) makes the interesting and salient point that it is difficult to actually define a "battlefield" in conflicts where battles took place over vast areas and which are often remembered and reenacted hundreds of miles from the original sites.

18. For example: Ryan, *Battlefield Tourism: History, Place and Interpretation;* David W. Lloyd, *Battlefield Tourism: Pilgrimage and Commemoration of the Great War in Britain, Australia and Canada 1919–1939.* Oxford, 1998; *Grief Tourism* website www.grief-tourism.com. Accessed December 2007; Neil Hanson, "Battlefield Tourism: Nothing New," *TravelMag.* 9 November 2005, available at www.travelmag.co.uk/article_938.shtml. Accessed December 2007.

19. For context on European neutrals in the 1930s see Albert Kersten, "Endangered Neutrality: The International Position of Luxembourg, Belgium and the Netherlands during the Interwar Years, 1919–1940," in Coenraad Tamse, Gilbert Trausch, eds., *Die Besiehungen zwischen den Niederlanden und Luxemburg*

im 19. und 20. Jahrhundert. Les relations entre les Pays-Bas et le Luxembourg aux XIXe et XXe siècles. The relations between the Netherlands and Luxembourg in the 19th and 20th centuries. The Hague, 1991, pp. 103–114; Neville Wylie, "Introduction: Victims or Actors? European Neutrals and Non-Belligerents, 1939–1945," in Neville Wylie, ed., *European Neutrals and Non-Belligerents in the Second World War.* Cambridge, 2002, pp. 1–30; Jukka Nevakivi, ed., *Neutrality in History. La neutralité dans l'histoire. Proceedings of the Conference on the History of Neutrality Organized in Helsinki 9–12 September 1992 under the Auspices of the Commission of History of International Relations.* Helsinki, 1993; Bob Moore, "The Posture of an Ostrich? Dutch Foreign Policy on the Eve of the Second World War," *Diplomacy and Statecraft.* 3, 3, November 1992, pp. 468–493.

20. The history of neutrality sits firmly within the fields of diplomatic, economic, international law, and, to some extent, military history. Notable exceptions include Abbenhuis, *The Art of Staying Neutral;* Hans A. Schmitt, *Neutral Europe Between War and Revolution 1917–1923.* Virginia, 1988; and scholarship on Switzerland and Ireland during the Second World War.

21. Hans Blom, "In de ban van goed en fout," Oration, December 1983; Chris van der Heijden, *Grijs Verleden. Nederland in de Tweede Wereldoorlog.* Amsterdam, 2003.

22. It is important to note here that Rosen is discussing works aimed at adults, not at children.

23. ANZAC stands for Australian and New Zealand Army Corps and refers to soldiers from Australia and New Zealand who were sent to fight for the British Empire during the First World War.

24. Joe Dante, dir., *Small Soldiers.* Universal Pictures. 1998.

25. The Dixie Chicks, "Easy Silence," *Taking the Long Way.* Sony. 2007.

26. Ibid., "Not Ready to Make Nice."

27. Tagline from the DVD release of Barbara Koppel, Cecilia Peck, dirs., *Shut Up and Sing.* Cabin Creek Films. 2007.

28. Andrea Liss, *Trespassing Through the Shadows: Memory, Photography and the Holocaust.* Minneapolis, MN, 1998, p. viii, cited in Wendy Kozol, "Domesticating NATO's War in Kosovo/a: (In)Visible Bodies and the Dilemma of Photojournalism," *Meridians: Feminism, Transnationalism, Race.* 4, 2, March/April 2004, pp. 1–38.

CHAPTER 1

Gilt Epaulettes for a Gilded Age: Citizen Volunteers and Martial Culture in Post–Civil War New York

Mark A. Potter

Explorations of war commemoration in the Gilded Age of the United States have suggested a growing public infatuation with military history.[1] Beginning in the early 1880s, veterans' reunions, memoirs, and popular histories of the war had begun to influence the development of America's patriotic culture. By the 1890s, Victorian Americans had eagerly read Civil War memoirs, fiction, and poetry; visited Civil War battlefields; and dedicated public soldier monuments. The legacy of America's martial past, in particular the American Civil War, heavily influenced developing ideals of citizenship and nationhood. War commemoration glorified the martial achievements of American manhood and offered Victorian men the opportunity to reaffirm their virility and vitality through patriotic connection with their martial forebears.[2] By the century's end, the cult of martial manhood in America was strong enough that hundreds of thousands of young men volunteered for service in the war with Spain in 1898. Most did not see active service, but the vehicles through which they volunteered were the state-based militia organizations that made up the nation's part-time military reserve, commonly known as the National Guard. In this chapter, I suggest that the National Guardsmen were an important but overlooked agent of late-nineteenth-century war commemoration and the popularization of martial ideals in the Gilded Age.

Of particular relevance to this collection is that the fascination with martial culture was a peacetime phenomenon—its development was largely supported by noncombatants, including those in the ranks of the Guard.

Apart from the veterans who joined Guard units after the Civil War and members of the few regiments that eventually saw service in 1898, the majority of Gilded Age Guardsmen never saw combat. Indeed, their knowledge of war came from listening to the stories of veterans, absorbing representations of war in popular culture, and participating in mock battles during their part-time military service. In the last decades of the century, increasing numbers of civilians with no direct military experience were drawn to service in the Guard. This trend reflected a growing fascination with warfare that emerged from Civil War commemoration and the fraternal and martial attractions of Guard service.

The process of establishing warfare at the center of American life began soon after the Civil War's end. Southern writers romanticized the Old South and celebrated the martial prowess of Confederate soldiers, a movement that became known as the Lost Cause.[3] The victorious North was slower to recognize the importance of its martial heritage, but from the early 1880s, veterans' reunions and commemorative events aimed at reconciling both North and South occurred with increasing frequency. While veterans may have disagreed on the rights and wrongs of the conflict, they found common ground in celebrating the martial prowess of American soldiers. Participation in these events gave the Guard a new national focus. Historians, however, have rarely looked at the Guard from this perspective. America's citizen soldiers have been widely discussed in terms of the failure or success of U.S. military policy, viewed with contempt in relation to their military effectiveness, or examined in their role as instruments of social control, particularly during the industrial troubles of the late 1870s.[4] These approaches have provided a convenient way to put the "national" into the National Guard.

The essential problem with making broad national claims about the Guard's influence is that Guard units were reflections of the local communities and state political structures that supported them. Jerry Cooper, a historian of the Guard, has noted that the "most significant problem in writing the history of the militia and the National Guard . . . is to combine the purely local, then colonial or state experience, with that of Imperial or National history."[5] Paradoxically, much work remains to be done on both their importance to local communities and their role in social, cultural, and political issues of national scope. While New York Guardsmen, for example, represented the often parochial interests of their state, they were also heavily involved in the organization of the National Rifle Association in the early 1870s and campaigned for recognition of the Guard as the nation's military reserve. Through their participation in commemorative events with a national focus, the martial pageantry of their displays, and their symbolic and military associations with the Civil War, Guardsmen were influential in maintaining the centrality of war among late-nineteenth-century Americans. The Guard's regular martial displays reinforced to an often enthusiastic public, which was steeped in the mythology of Americans at war, the view that the exemplar of martial manhood was the citizen volunteer, a view sustained through nationwide war commemoration.[6]

Alongside the dedicating of public soldiers' monuments, writing memoirs, and joining veterans' organizations like the Grand Army of the Republic (GAR), veterans supported commemoration through reunions. Nonveteran Guardsmen and civilians were often present at these events. One of the earliest recorded reunions of Northern and Southern veterans occurred in 1881 when members of the 71st Regiment of the New York National Guard, the focus of this chapter, traveled to New Orleans to celebrate Mardi Gras with Southern veterans and militia. The historian Paul Buck, in his 1937 work *The Road to Reunion,* notes that the visit of the 71st was the forerunner of the reunions of the 1880s and 1890s, which culminated "in two great spectacles," the twenty-fifth anniversary of the battle of Gettysburg in 1888 and the dedication of the national military park at Chickamauga and Chattanooga in 1895.[7] The importance of this statement is that the experience of the 71st was not an isolated one and should be viewed against the broader background of a developing culture of commemoration and romantic sentiment about warfare, in which Guard units across the nation took part in events similar to those described here.

The 71st was formed in the 1850s as a volunteer militia unit in New York City and fought in the first major engagement of the Civil War, at First Bull Run in 1861. For the remainder of the war, it served largely as a home guard force, although many of its members volunteered for duty in the Union army. In post–Civil War America, citizen volunteers were broadly known as National Guards although some states and many commentators still called them militia. New York, in fact, formalized the term "New York State National Guard" during the war, and the 71st became part of the official civil-military apparatus of the state. As part of the Federal expeditionary force sent to Cuba in 1898, it was the only Guard unit to fight at San Juan Hill alongside regiments of African American regulars and Teddy Roosevelt's famous Rough Riders.[8]

This chapter, however, looks beyond the battle history of the regiment, placing it, and the Guard more broadly, within a framework of war commemoration, to explore the development of martial culture in the United States in the late nineteenth century. It examines the impact of war upon noncombatants: on the young Guardsmen who would never see battle but who were in day-to-day contact with veterans of the Civil War; on their friends and families who socialized with them at Guard encampments; and on the public who became increasingly enthusiastic about their martial displays as the mood of commemoration, reunion, and adoration of America's military past took hold in the 1880s and 1890s. Looking through the lens of a unit of citizen volunteers, a complex process of negotiated cultural memories is revealed, whereby regimental traditions of volunteering and Civil War service are mixed with a developing national patriotic martial culture.

Of broader relevance to this collection is that the celebration of a romanticized martial past maintained the centrality of war for Victorian Americans and, by instilling martial ideals in civilian society, supported the rise of an industrialized and militarized United States after the Civil War. The martial enthusiasm behind the imperial adventures of the turn of the century has been seen as an immediate result of the late-nineteenth-century crisis of masculinity, whereby war was seen as a panacea to male effeminacy.[9] Theodore Roosevelt warned that the "greatest danger that a long period of profound peace offers to a nation is that of [creating] effeminate tendencies in young men."[10] In response, Roosevelt and many other Victorian men looked for models of independent, vigorous manhood. The martial valor of Civil War soldiers was a perfect fit. In this chapter I suggest an alternative view that the martial enthusiasm that supported America's quest for empire in 1898 was the culmination of decades of war commemoration and a developing cultural identification of national progress with the cultivation of martial ideals. This chapter contributes to our understanding of the ways in which societies utilize the martial enthusiasm of noncombatants in order to organize themselves for conflict and encourage their youth to volunteer for war, which would become a key factor in the ability of western nations to fight the total wars of the twentieth century.

Commemoration

The American novelist Carl Sandburg, writing of his war service in 1898, remembered that over all of the volunteers in the war with Spain "was the shadow of the Civil War and the men who fought it to the end that had come only 33 years before our enlistment."[11] Sandburg's observation reminds us that Victorian America was essentially a postwar world, in which the sacrifices of a highly destructive Civil War were still keenly felt and the martial and national legacy of the conflict weighed heavily on the minds of postwar generations. Eugene S. Eunson, a major of the 71st, noted the difficulties facing the Guard in the years immediately after the Civil War. There was a reaction against the "militia" so that "in the popular estimation it was almost a discreditable thing to be a member."[12] None of the Guard organizations in New York in the late 1860s were at full strength, and only one regiment, the 7th, had quarters that met the basic requirements of a regimental armory.[13] Yet, as memories of the violence dimmed and veterans began to publish memoirs of their war service, volunteering regained some of its prewar appeal.

Studies of commemoration have concentrated mainly on the remembrance activities of veterans and the organizations that spoke for them, the Grand Army of the Republic, for example.[14] These studies stress that in the last two decades of the nineteenth century, in a spirit of militaristic nationalism, veterans from both North and South trod the path of reunion, glorifying and sentimentalizing the Civil War, and celebrating the prowess of the American soldier. Much of the history of sectional reunion concentrates on how the South came to terms with defeat, famously of course through the mythology of the Lost Cause, which romanticized the Old South. In the 1990s, however, historians such as Nina Silber, David W. Blight, and Kirk Savage sought to reassess how the North came to terms with victory and how this contributed to the reunion process. David W. Blight, in *Race and Reunion,* argued effectively that sectional harmony and the martial valor of white soldiers emerged as a dominant motif of commemoration, veterans' reunions, and memorial celebrations in the 1880s and 1890s. This was at the expense of competing narratives like those of slavery and emancipation.[15]

The participation of Guard units in these commemorative activities has generally been overlooked. Almost twenty years after they fought for the Union at First Bull Run, the 71st New York traveled to New Orleans at the invitation of Southern veterans and community leaders to participate in the Mardi Gras celebrations of 1881. Militia units from New Orleans, the Washington Artillery, for example, had also been present at First Bull Run. The journey was a chance for the 71st to confirm its martial reputation, something to which the regiment clung tenaciously in the face of growing Lost Cause celebration of Southern martial superiority, the tendency for the North to celebrate only its great victories, notably at Gettysburg, and the recruiting and financial difficulties faced by Guard formations in the immediate postwar environment. The establishment of a martial reputation during the Civil War, always a source of regimental and local pride, took on greater significance as the process of reunion and reconciliation took hold in the 1880s.

The journey came at a time of increasing interest in New York in remembering the Civil War and in the preservation of its heritage. In 1881, the *New York Times* noted that the "relics and records of the war have been too long neglected. It is high time that they were arranged and protected with a . . . reverent care worthy of the patriotism of our volunteer soldiery, worthy of the great State of New York."[16] The 71st's journey, therefore, came at a pivotal point on the path of sectional reunion and at a time of increasing public and government interest in the preservation of New York's Civil War heritage. The members of the 71st were invited to New Orleans as guests of both the Louisiana State Militia and leading members of New Orleans' society. The contingent consisted of two hundred rank-and-file volunteers along with five veterans of Bull Run. The majority of the touring party were young Guardsmen with no direct experience of war. While taking note of the broader social and cultural environment of the journey, it is the interaction of the veterans and their younger charges that is valuable in investigating the way in which regimental traditions and martial ideals of manhood were passed from one generation to another. This incident from the journey is an excellent illustration of this intergenerational legacy.

At Hammond, Louisiana, a detachment of the Washington Artillery welcomed the regiment with a 71-gun salute. Each man in the detachment was a veteran. Gentlemen of business and social prominence, they proceeded to exchange war stories with the veterans of the 71st.[17] One of the artillerymen said that he had a Yankee bullet in him somewhere, which he would like to present to a Northern veteran if he could only get it out. A member of the 71st, with tears in his eyes, declared that he had a piece of "Johnny" bullet in his thigh "and that he would give a thousand dollars to get both of them out so they could exchange. Then the two veterans went off to get a drink, and everybody within hearing, wished that they had a bullet in some portion of their body."[18] Though amusing, this account of the exchange of war stories is an important example of the way in which veterans imbued younger nonveteran members of the regiment with traditions and memories that would shape their views on reconciliation and warfare.

Another significant event was attendance by the regiment at a ceremony to honor the Confederate dead. Arriving at Greenwood Cemetery, near the place where Andrew Jackson turned back the British in 1815, the regimental band played "Nearer My God to Thee" accompanied by the voices of several thousand spectators. The men of the regiment with reversed arms walked slowly around the Soldier's Monument. The chaplain of the 71st regiment, Dr. Martyn, declared that generals Ulysses S. Grant and Robert E. Lee, Stonewall Jackson, and William Tecumseh Sherman would join the pantheon of American patriotism as would the host of men who fought on one side for the National idea and on the other for the Lost Cause.[19] The use of "national idea," that is, of one nation undivided, the Union, is instructive. On the one hand, the regiment was celebrating the martial prowess of soldiers from New York City, yet, on the other, their celebration is imbued with the language of patriotism and nationhood. This represents the formation of a national patriotic culture through commemoration of shared martial sacrifice.

Here then is an early commemorative event, in which both veterans and younger Guardsmen took part, which gave meaning to the sacrifices of the Civil War generation. The ceremony foreshadowed the many commemorations in the coming decades that stressed the brotherhood of all American soldiers regardless of the cause for which they fought. A Southern newspaper, the *New Orleans Democrat,* noted the significance of Northern soldiers decorating the graves of Confederate dead at a place where Americans had earlier defended their nation against the British. We "know that the act will be productive of great good—proving to the Southern people that the men of the North are as magnanimous in peace as they found them brave in war."[20] Thus the salute both affirmed the mutual bonds of soldierly sacrifice and valor essential to reunion and confirmed the older national bonds of the citizen soldiery, bonds established in revolution, which were confirmed through the mythology of a war of national defense. The expedition of the 71st to New Orleans has significance beyond Buck's assessment of it and beyond its place as one of the first veteran reunions. This was an active National Guard unit that included some veterans in its ranks but with the rank and file made up of nonveterans aged in their twenties and thirties. The veterans were handing down to non-veterans not just regimental traditions but also a particular remembrance of the war, shared through their stories and their interactions with their Southern counterparts.

One final aspect of Guard commemorative activities to consider is their regular attendance at summer training encampments. Between 1881 and 1892, every state revised its military code to establish organized, voluntary National Guard formations.[21] These military codes mandated that Guard units undertake regular training in the field. Summer training camps became commonplace and states petitioned the federal government for inspections and training by regular army officers. By 1887, 13 state summer camps received inspections by regulars. The philosophy behind sending Guardsmen to summer camps was not to train them to be able to suppress internal disorder but rather to ensure that the Guard was a well-equipped and disciplined organization ready to take the field as the national reserve force.[22] In doing so, the National Guard became a key part of United States military planning and in the process embedded the culture of military prowess in society. As a result, the prestige of the National Guard increased and young men who volunteered for duty in it represented the "ideal citizen" in every sense.

While the most vital military function of state encampments was teaching individual units how to operate together in a simulated battlefield environment, military periodicals espoused the moral and physical benefits of Guardsmen training together. Encampments promoted discipline and drill and imparted "a better knowledge of the requisite details of a soldier's life, than whole seasons of indoor instruction and practice in armories and arsenals."[23] Others noted that if Guardsmen from various states trained together, it would have national effects and "allay sectional prejudice, create new friendships, and weld the young soldiers of the Republic together indissolubly."[24] Thus, like participation in commemorative events, encampments encouraged a national focus among Guard units. Yet encampments took the civilian-military association much further than commemorations through the performance of war during mock battles and the attendance of civilians at these events. Mock battles allowed peacetime indulgence in shooting and fighting that was as close as the Guardsmen would get to the descriptions of war they read about in Civil War histories and memoirs. Encampments of both Guardsmen and veterans offered young men the opportunity to take the field with the heroes of the war and thus further impart the martial traditions established in conflict.

Encampments were also sites of commemoration, where young Guardsmen could experience a taste of military life while imagining that they were on a Civil War battlefield. The presence of veterans in the ranks and members of GAR sustained these imaginings. GAR posts sometimes camped with Guardsmen. At the New Jersey department encampments of 1878, 1881, and 1883, veterans engaged in sham battles with New Jersey Guardsmen. GAR posts also engaged in shooting matches with Guardsmen.[25] The reenactment of battle at GAR camps and the sham battles at Guard encampments transmitted an interpretation of the past, their "communicative performance [providing] a dramatic vehicle for making rememberings in common possible."[26] Thus, encampments and battles were inducements to memory in which the participants and, to some extent, the observers had the momentary experience of being "in the past."

This experience of being "in the past" could also be transmitted to observers. The proximity of encampments to major cities meant that they were in easy traveling distance of friends and family members of the Guardsmen. While friends and family would travel considerable distances to visit soldiers in camp, they did not need to go so far afield to witness martial displays of a military nature. When the city of New York took over responsibility for Van Cortlandt Park, north of Brooklyn, in the 1880s, it turned part of the park into a parade ground. While curlers and skaters used the park's lake in the winter months, the park grounds also afforded excellent opportunity for field exercises in warmer weather. Over a hundred acres of land could accommodate the maneuvers of thousands of men. The *Army and Navy Journal* proclaimed that the "opportunities afforded by this extensive parade ground . . . for military exercise on a scale unprecedented in this city or its vicinity will attract tens of thousands of spectators on special occasions." Hills overlooking the training area on its northern and western sides enabled "over a hundred thousand spectators" to see a "brilliant spectacle as infantry, cavalry and artillery go through their exercises or arranged in mimic battle."[27]

Military spectacle such as this had a definite effect on both the reputation of the Guard and attitudes toward war and martial ideals. In 1887, the year of the Centennial of the Constitution, celebrations and parades were held across the country. The martial display that accompanied these commemorative events had a significant impact on public attitudes toward the military and in particular toward citizen soldiers. The *Army and Navy Journal* noted that citizens who had regarded citizen soldiers with indifference or contempt in the past "were so impressed that they made a complete change about" and were converted to the belief that "it is the citizen soldiery that is the greatest safeguard of the country or of the State in any emergency."[28] In Philadelphia, state militia and Guard organizations, regular army units, and veterans from both North and South were reviewed by President Grover Cleveland. The Philadelphia press reported that there "was something in the spectacle of the inspiring military display . . . that aroused and impressed upon the people not only the thought that Americans are a martial nation, but that a certain amount of martial training or service . . . is one of the duties of citizenship." Furthermore, the display "had the effect of stirring up the military feeling and enthusiasm of thousands as nothing else could have done save a genuine call to arms."[29]

Studies of Civil War commemoration have effectively dealt with the way in which the memories and commemorative activities of veterans romanticized war, leading to reverence of the volunteer soldier. This is only part of the story, however, albeit an important one. The veterans had fought their war. Their sons and grandsons had to find their own tests. How were the ideals of martial manhood passed down through these generations? It was done partially through popular remembrance of the war and commemorative events like the New Orleans visit. But it was also sustained by the regular appearance of Guard units on the streets of New York and other major cities.

The Guard on Parade

The martial displays Guardsmen witnessed as boys and youths and participated in as young men promoted virtues of patriotism, organization, discipline, skill with weapons, fraternity, and masculinity. Displays of martial manhood by National Guard units were regular events. Increasing interest in Guard membership, with its associated display and rituals, reflected a more general tendency toward joining clubs and associations, in part a reaction to the growth and anonymity of the city, providing a haven from both domestic life and the stresses of the modern city.[30] Parading was also a reaction to this change. Fraternal organizations, unions, temperance societies, and ethnic associations were among the many that regularly paraded. In 1888, *Harper's Weekly* noted that the militia "differ in degree, though not in kind, from those orders, for keeping secrets, or for encouraging a distaste for strong drink, which also wear bright and attractive regalia, and go about in processions, with banners and music, and a pomp that cannot be distinguished at a distance from real war."[31] However, interest in martial display coincided with a wider concern in the late nineteenth century for the virility of the American people and an emerging interest in history and warfare.

When the 71st returned from New Orleans to New York in 1881, the reception it received illustrates the powerful patriotic associations that could be created in the minds of spectators observing such events. The *Army and Navy Journal* reported that the men, faces bronzed, appeared like veterans and reminded onlookers of "days long past, when the heroes of many a hard fought field marched up Broadway on their return from the war." The reception they received on their return from the South indicated that their mission, " to cement the friendship of the soldiers, North and South, was most fully appreciated by the people of New York City."[32] The regiment's return was heavily steeped in the symbolism of war, echoing the return of the victorious Union regiments in 1865. Yet it was also the celebration of the beginnings of the reunion between North and South and hence was as a force of regional and national progress. The ability to arouse people's patriotism and, thus, affect attitudes toward other sections of the nation was a potent force of reunion sentiment and popularization of martial culture.

The martial displays of citizen soldiers—parades, shooting contests, mock or sham battles, balls and concerts in regimental armories, and participation in commemorative activities—connected the manly cultivation of martial ideals with national progress. Their uniforms and displays of skill with weapons, while presenting an image of powerful manhood also reflected a psychological need to be "in control" both physically and morally. This need extended to technology, symbolizing their individuality and autonomy from market and social forces as well as their mastery of modern weaponry; their bodies, thus showing the manly discipline required for success in life; and the social space of the streets they paraded in, which helped to define the boundaries of citizenship and public order. As public spaces, the main thoroughfares of New York provided not only corridors for transportation but were also, as Mona Domosh has noted, "sites for the displays of social class and political power." Because it extended the whole length of Manhattan, Broadway was the "grand boulevard of display."[33]

The 71st marched along Broadway regularly, going to and returning from war in 1861 and 1898, on their regular intrastate and interstate visits to participate in reunions and Guard training exercises, but most often to participate in parades during local and national days of celebration and commemoration. While not many cities could compete with Broadway in terms of its ability to show off parading Guardsmen to thousands of spectators, parades played a significant role in the ceremonial and celebrative life of most American cities. Indeed, parades were a primary form of public entertainment in the nineteenth century.[34] On festive occasions, holidays, and during important local, state, and national events, regiments, brigades, and divisions of National Guardsmen and militia would march through the streets in front of large crowds.

After a parade by the 71st New York Guardsmen in 1871, the *Army and Navy Journal* noted that the regiment was "the most manly organization in the National Guard; scarcely a member is without a beard; and as they appeared on this occasion in their full dress, they well deserved and certainly gained the admiration of all observers."[35] While it was not unusual for a military periodical to describe a unit in this way, the description is significant in defining the regiment's overt signs of manhood: fine parading, facial hair, and full-dress uniforms. While the uniform held great importance to the wearer, its significance stems as much from the symbolic associations it had for the audience. Through a process of myth-making and abstraction, uniforms became a system of "sartorial codes." These codes functioned as a "vocabulary of stereotypes" by which the observer conceptualized their world. Uniforms then were overtly political, visually confirming perceptions of state authority, national military strength, and the martial prowess of the American male.[36] In the 1880s, the state of New York adopted a regulation blue uniform for all its Guard units. The sight of the Guard parading in blue had patriotic connotations, blue being the color of the uniforms of the victorious Union army. As well as presenting an image of a unified state Guard, the sight of soldiers parading in blue uniforms carried powerful associations of martial prowess and national progress, and thus were an effective public display of both patriotism and nationalism.

Martial spectacle was also a vital recruiting tool, with much competition between the city's Guard organizations for the best young men to join their ranks. Importantly, it was the masculine prestige and sexual attractiveness attached to being a Guard, as much as a sense of patriotism, that drew many young men to volunteer for service. When the 71st traveled to New Orleans, they stopped in Cincinnati en route. The marching of the unit "in perfect order" down Fourth Street under the "glances of the hundreds of Ohio ladies," the men's blue coats, red blankets, and shiny Remington breechloaders, presented a brilliant appearance that "will long be remembered."[37] The regiment's martial bearing and shiny rifles symbolized American martial prowess and indeed the virility of the men themselves. The uniforms of the Guardsmen helped to underline "socially defined expectations for behaviour,"[38] which was both gendered and conformist and reinforced links between militarism, masculinity, and citizenship. The uniforms and disciplined display of young men bonded by the fraternity of the regiment and the patriotic ties of service exemplified, as Susan Davis has noted, "all a male citizen should hope to be."[39]

The parades of Guard organizations, especially when connected to commemorative events, contributed to the formation of a national patriotic culture. War commemoration became intimately connected to the portrayal of martial manhood and the emerging connection of martial ideals with nationalism that would reach its fullest expression with the national reconciliation engendered by war with Spain in 1898. That the Civil War had a deepseated impact upon American attitudes toward war and ideals of manhood is evident in the attitudes of volunteers in 1898, which were conditioned by indoctrination into the martial culture of the United States. The promotion of martial manhood was assisted not only by war commemoration but also by a general interest in war and violence as the century drew to a close.[40] Games depicting war were enjoyed by family members of all ages, apparently untroubled by grisly images of bloodshed. The popularity of board games inspired by war reflected a late-nineteenth-century fascination with violence and with current events and foreign affairs. In 1895, Parker Brothers produced the "Game of Napoleon: The Little Corporal," while "Mimic War" contained a box of 30 military figures with a cover depicting the Franco-Prussian conflict, although the figures were in the costumes of 1898.[41]

Thus, before the real war in 1898, there were many fantasy wars played out by boys, youths, and military-minded men. They were played out in mock battles at Guard encampments, through reading newspapers and military and popular periodicals with their description of the world's armies and navies, regular comparisons of the strengths of other countries' forces and fictional accounts of future battles, and in the continued popular interest in reminiscences of warfare. Carl Sandburg remembered the importance of history books in his early education. He read J. T. Headley's *Napoleon and His Marshals* and John Abbot's *The History of Napoleon Bonaparte* to "see what kind of fighter he was." His favorite history books, however, were the series by Charles Carleton Coffins. Coffins's *The Boys of '76* made Sandburg feel like he "could have been a boy in the days of George Washington and watched him on a horse, a good rider sitting easy and straight, at the head of a line of ragged soldiers with shotguns."[42] Sandburg and his childhood friends were indoctrinated by popular culture, marching veterans, and Guardsmen into the masculine affairs of adventure and warfare.

The growing interest in the possibility of a major war in the 1890s was noted by the literary journal *The Bookman*. The journal noted that there was "no more striking proof" of the drift toward war than "the extraordinary amount of space devoted by editors and publishers to the discussion of military themes." However, the "feverishness of public sentiment and the growing interest in whatever relates to battle" were not confined to the United States, but were common to all the "Western peoples." The writer's justification for this goes right to the heart of the appeal of warfare to young men seeking to test their manhood. War was the only game "that can thrill the nerves and give the fullest play to the emotions."[43] The *Bookman* succinctly describes the process of remembering and forgetting that had taken place during the three decades that had passed since the end of the Civil War.

Describing New York National Guardsmen going to their encampments in May of 1898, Vaughan Kester notes the responsibility felt by the sons to accept the mantle handed down by veterans. "A new generation had arisen since the Civil War. The battles, if they came, were to be fought by the sons of those who had fought before . . . The musket and the sword had passed from sire to son." Kester revealed that while all were "eager for the actual scenes and experiences of war and the grim reality of battles," it was through no "personal love of strife or conflict." They were motivated by "an eagerness to prove their worth, a desire to pass the test, to escape from the last doubt, to become indeed tried soldiers."[44] Volunteers believed they owed a debt to the Civil War generation and envied their place in history. Private Charles Johnson Post of the 71st, recalled the regiment's return from Cuba in 1898:

> I could imagine it as it was in the days of Sherman and Grant and Lincoln, and see that last parade of the "Boys in Blue," with the Civil War that had closed but thirty years before. I could see the ranks, tanned and grizzled, of veterans of great battles, and rugged campaigns that had made history, men who faced death and survived. I envied those veterans their memories, and their great parade.[45]

Post was only 24 when he volunteered for service in 1898, so he could only imagine the grand parades at the Civil War's end. However, as a young man growing up in New York, he would regularly have witnessed many parades and other displays of a martial nature. The sight of marching men, veterans and Guardsmen alike, in blue uniforms was marked on Post's mind as symbolic of American military prowess and martial manhood.

A. Maurice Low in an article on the volunteers of 1898 noted that while "it is easy to create armies on paper, soldiers, unlike poets, are not born, but made, and . . . the process of manufacture is a slow and difficult one."[46] The process by which large numbers of Americans from many sections and communities accepted martial ideals of manhood and supported war in 1898 has been widely documented. However, the Guard's role in that process has not been. The 71st New York played a pivotal role in the dissemination of martial ideals to the wider community of New York City and surrounding states throughout the postwar period. While displays were often on a smaller scale than those presented by the 15,000 members of the New York Guard, they were regular events that promoted nationalism, sectional reconciliation, the military power of the State, and ideals of manhood.

Conclusion

Efforts to improve the martial efficiency of Guard organizations in the late nineteenth century paralleled calls for increased efficiency in society in general and were in effect designed to make the "best" soldiers from the "best" citizens.[47] Military academies, the growth of professionalism in sport and business, and the multiplication of expert professions all point to the increasing significance of technical expertise, both in the military and in Gilded Age civilian society.[48] In short, bureaucratic techniques of organization, rationality, and advances in weaponry and logistics were applied to the making of war. Indeed, as powerful as long-range weaponry was becoming, "it was the social technique of bureaucratically rationalized violence" that enabled Western nations to succeed in their colonial ambitions.[49] Personal heroism seemingly had little to offer in the face of industrialized warfare. Yet, the imagery of masculine heroism, promoted through war commemoration and the martial displays of Guardsmen, had important cultural and organizational effects, in terms of preparing men for war and providing them with cohesion in battle.

The patriotic response of citizen soldiers in 1898 to the war with Spain was not just a reaction to the immediate social conditions of the 1890s. It was the culmination of decades of war commemoration and a developing cultural identification of national progress with the manly cultivation of martial ideals during peacetime. Part of this process was the incubation of martial ideals in the National Guard and their propagation to the wider public through regular martial displays from the end of the Civil War. The image of the American citizen soldier as heroic warrior would be utilized in the coming century to raise armies and fight modern total wars. The promotion of martial manhood through commemoration and martial display was a powerful social force that incubated martial ideals of citizenship and national progress through warfare. It promoted nationalist patriotism among noncombatants and was a significant factor in preparing the nation psychologically for both reunion between North and South and popular involvement in the Spanish-American War. The National Guard, therefore, played a key part in promoting the culture of war in Gilded Age United States and in reinforcing traditional ideas about men as warriors and soldiers.

Notes

1. The Gilded Age is the period between the assassination of Abraham Lincoln in 1865 and 1901. It is associated with huge growth of wealth, leaps in industrial development, and westward expansion, as well as accompanying issues of governance and corruption. Sean Dennis Cashman, *America in the Gilded Age from the Death of Lincoln to the Rise of Theodore Roosevelt.* 3rd ed. New York, 1993.
2. Anne C. Rose, *Victorian America and the Civil War.* New York, 1992, pp. 245–247. Rose examines the relationship between Victorian culture and the Civil War, arguing that while war commemoration was influenced by and "strengthened a broader mood of cultural criticism that centered on similar attention to warfare and history," it also offered Victorian Americans "avenues of emotional escape" from the effects of immigration, urbanization, and industrialization through a romanticized past. For an alternative view, see Charles Royster, *The Destructive War: William Tecumseh Sherman, Stonewall Jackson and the Americans.* New York, 1991.
3. Gaines M. Foster, *Ghosts of the Confederacy: Defeat, the Lost Cause and the Emergence of the New South, 1865–1913.* New York, 1987.
4. Jerry Cooper, *The Militia and the National Guard Since Colonial Times: A Research Guide.* Westport, CN, 1993, p. 2. The classic critique of the militia's military effectiveness is Emory Upton, *Military Policy of the United States.* Washington, D.C., 1904. For a contemporaneous view in support of the citizen soldier, see Grand Army of the Republic commander John A. Logan, *The Volunteer Soldier of America.* Chicago, 1887. Also see Russell F. Weigley, *Towards and American Army: Military thought from Washington to Marshall.* New York, 1962, pp. 137–161; Russell F. Weigley, *History of the United States Army.* Bloomington, IN, 1984; Allan Millett and Peter Maslowski, *For the Common Defense: A Military History of the United States of America.* New York, 1984; Jerry Cooper, *The Rise of the National Guard: The Evolution of the American Militia, 1865–1920.* Lincoln, NE, 1997; Jim Dan Hill, *The Minute Man in Peace and War: A History of the National Guard.* Harrisburg, PA, 1964; John K. Mahon, *History of the Militia and the National Guard.* New York, 1983. For state-specific studies, see Cooper, *Rise of the National Guard;* John K. Mahon, "Bibliographic Essay on Research into the History of the Militia and the National Guard," *Military Affairs.* 48, April 1984, pp. 74–77.
5. Cooper, *Rise of the National Guard,* p. 2.

6. Other authors have used terms similar to "martial manhood" to discuss the social importance of late-nineteenth-century martial ideals: Cecilia O'Leary, *To Die For: the Paradox of American Patriotism*. Princeton, NJ, 1999, pp. 30–33; T. Jackson Lears, *No Place of Grace: Anti-Modernism and the Transformation of American Culture, 1880–1920*. New York, 1981, pp. 98–102; Mark Kann, *On the Man Question: Gender and Civic Virtue in America*. Philadelphia, PA, 1991, pp. 15–19; and E. Anthony Rotundo, *American Manhood: Transformations in Masculinity from the Revolution to the Modern Era*. New York, 1993, pp. 232–239.

7. Paul Buck, *The Road to Reunion, 1865–1900*. Boston, 1937, p. 257.

8. For a more detailed view of the 71st, see Mark Potter, "A Good Soldier, a Good Shot and a Good Fellow: The Seventy-First New York, Martial Manhood and the Shadows of Civil War, 1850–1898" PhD thesis, University of Melbourne, 2005. Also see Fred L. Israel, "New York's Citizen Soldiers: The Militia and Their Armories," *New York History*. 42, April 1961, pp. 145–156; John F. and Kathleen Smith Kutolowski, "Commissions and Canvasses: The Militia and Politics in Western New York, 1800–1845," *New York History* 63, 1, 1982, pp. 4–38; Ronald Howard Kotlik, "Fixed Bayonets: The New York State National Guard during the Era of Industrial Unrest, 1877–1898" PhD thesis, State University of New York, 2005; and Russell S. Gilmore, "New York Target Companies: Informal Military Societies in a Nineteenth-Century Metropolis," *Military Collector and Historian*. Summer 1983, pp. 60–66.

9. David Greenberg, for example, writes that male effeminacy as a perceived collective phenomenon emerged in Britain around the same time (*The Construction of Homosexuality*. Chicago, 1988, pp. 383–393). Also see Robert Nye, ed., *Sexuality*. Oxford, 1999; George L. Mosse, *The Image of Man: The Creation of Modern Masculinity*. Oxford, 1996, especially p. 15.

10. Cited in Greenberg, *Construction of Homosexuality*, p. 393.

11. Carl Sandburg, *Always the Young Strangers*. New York, 1952, p. 409. Also see Stuart McConnell, *Glorious Contentment: The Grand Army of the Republic, 1865–1900*. Chapel Hill, NC, 1992, p. xii; Rose, *Victorian America and the Civil War*, p. 245; Michael S. Sherry, *In the Shadow of War: The United States Since the 1930's*. New Haven, CT, 1995, p. 1; Richard Slotkin, *Gunfighter Nation: The Myth of the Frontier in Twentieth-Century America*. New York, 1992, p. 89; and George M. Frederickson, *The Inner Civil War: Northern Intellectuals and the Crisis of the Union*. New York, 1965.

12. Eugene S. Eunson to Augustus Francis, in Augustus Theodore Francis, *History of the 71st Regiment, N.G.N.Y.* New York, 1919, p. 851.

13. Francis, *History of the 71st Regiment, N.G.N.Y.*, pp. 322–23.

14. McConnell, *Glorious Contentment*.

15. David W. Blight, *Race and Reunion: The Civil War in American Memory*. Cambridge, MA, 2001, pp. 1–5; Nina Silber, *The Romance of Reunion: Northerners and the South, 1865–1900*. Chapel Hill, NC, 1993, pp. 92–123; and Kirk Savage, *Standing Soldiers, Kneeling Slaves: Race, War, and Monument in Nineteenth-Century America*. Princeton, NJ, 1997, pp. 162–208.

16. "New York's Military Memorials," *New York Times*. 16 February 1881.
17. "The 71st in New Orleans," *New York Times*. 28 February 1881.
18. John F. Cowan, *A New Invasion of the South: Being a Narrative of the Expedition of the Seventy-First Infantry, National Guard, through the Southern States, to New Orleans*. New York, 1881, p. 41.
19. "The Seventy-First's Trip: Honoring the Confederate Dead," *New York Times*. 4 March 1881.
20. *New Orleans Democrat*. Cited in Cowan, *New Invasion of the South*. Appendices, p. 20.
21. Martha Derthick, *The National Guard in Politics*. Cambridge, MA, 1965, p. 16.
22. Adjutant General, *Annual Reports of the Adjutant-General of the State of New York*. Albany, 1874, p. 9.
23. "Encampments," *Army and Navy Journal*. September 1870.
24. A. C. Sharpe, "Organization and Training of a National Reserve for Military Service," *Journal of the Military Services Institute*. (JMSI) 10, 1889, p. 28. Also see Major Howard A. Giddings, "How to Improve the Condition and Efficiency of the National Guard," JMSI 21, 1897.
25. McConnell, *Glorious Contentment*, pp. xii, 47.
26. O'Leary, *To Die For*, p. 197. O'Leary talks specifically of veterans reenactments in the early twentieth century, but her conclusions are equally valid here. Also see Greg Dening, *Performances*. Melbourne, 1996.
27. "New York's Parade Ground," *Army and Navy Journal*. November 1887, p. 349.
28. "Pennsylvania National Guard," *Pennsylvania Times*. Cited in *Army and Navy Journal*. October 1887, p. 251.
29. *Army and Navy Journal*. October 1887, p. 251.
30. Arthur M. Schlesinger, "Biography of a Nation of Joiners," *American Historical Review*. October 1944. pp. 1–25.
31. "Value of Military Display," *Army and Navy Journal*. September 1888, p. 15.
32. "Seventy-first New York," *Army and Navy Journal*. March 1881.
33. Mona Domosh, "Those 'Gorgeous Incongruities': Polite Politics and Public Space on the Streets of Nineteenth-Century New York City," *Annals of the Association of American Geographers*. 88, 2, June 1998, pp. 209–226.
34. Susan G. Davis, *Parades and Power: Street Theatre in Nineteenth Century Philadelphia*. Philadelphia, 1986; Mary P. Ryan, *Civic Wars: Democracy and Public Life in the American City during the Nineteenth Century*. Berkeley, 1997. For a discussion of festivals as "arenas where factions struggled to define national identity and set the limits of citizenship," see Clifton Hood, "An Unusable Past: Urban Elites, New York City's Evacuation Day, and the Transformations of Memory Culture," *Journal of Social History*. 37, 4, 2004, pp. 883–913. Also see David Waldstreicher, *In the Midst of Perpetual Fetes: The Making of American Nationalism, 1776–1820*. Chapel Hill, NC, 1997; Simon P. Newman, *Parades and the Politics of the Street: Festive Culture in the Early American Republic*. Philadelphia, 1997.
35. *Army and Navy Journal*. 1 April 1871.

36. Nathan Joseph, *Uniforms and Nonuniforms: Communication through Clothing.* New York, 1986, pp. 102–111.

37. *Cincinnati Commercial.* Cited in Cowan, Appendices, p. 17. On the sexual allure of uniforms, see Valerie Steele, *Fetish: Fashion, Sex and Power.* New York, 1996, pp. 180–182; Tim Newark, *Brassey's Book of Uniforms.* London, 1998, p. 42.

38. Scott Hughes Myerly, *British Military Spectacle: From the Napoleonic Wars through the Crimea.* Cambridge, MA, 1996, pp. 58–59; Ruth P. Rubenstein, *Dress Codes: Meanings and Messages in American Culture.* Boulder, CO, 1995, p. 3.

39. Davis, *Parades and Power,* p. 71. For the British context, see Myerly, *British Military* Spectacle. For the use of military spectacle as a force of social control, see Karl Liebknecht, *Militarism and Anti-Militarism.* New York, 1972.

40. Thomas C. Leonard, *Above the Battle: War-Making in America from Appomattox to Versailles.* New York, 1978.

41. The games described were part of a display called "Mimic War" at the New York Historical Society, 2002. Games included "Game of Napoleon: The Little Corporal" (Parker Brothers, 1895), "Rough Rider Ten Pins" (R. Bliss Manufacturing Company, 1898), "Roosevelt at San Juan" (Chaffee & Seldon, 1899), "The Great Game: Uncle Sam at War with Spain" (Rhode Island Game Company, 1898), and "Game of the Little Volunteer" (McLoughlin Brothers, 1898). I have used the accompanying commentary as a source for this perspective.

42. Sandburg, *Always the Young Strangers,* p. 115.

43. "The Drift Towards War," *The Bookman: A Literary Journal.* April 1896, p. 154.

44. Vaughan Kester, "Transformation of Citizen into Soldier," *Cosmopolitan.* June 1898, p. 150–151. Also see Gerald Linderman, *The Mirror of War: American Society and the Spanish-American War.* Ann Arbor, MI, 1974.

45. Charles Johnson Post, *The Little War of Private Post.* Boston, 1960, p. 35.

46. A. Maurice Low, "Amateurs in War," *The Forum* 26, October 1898, pp. 157–166.

47. Linderman, *Mirror of War,* pp. 90–100.

48. The increasing number of military periodicals in this period is a good example of this process. Also see John P. Mallan, "The Warrior Critique of the Business Civilisation," *American Quarterly.* Fall 1956, pp. 216–230; Terry Mulcaire, "Progressive Visions of War in the Red Badge of Courage and the Principles of Scientific Management," *American Quarterly.* 43, 1, 1991, pp. 46–72.

49. R.W. Connell, *Masculinities.* Berkeley, CA, 1995, p. 192.

CHAPTER 2

Patterns from the Guardians of Neutrality: Women Social Democrats in Sweden and Their Resistance against Civil Defense, 1939–1940

Irene Andersson

The Swedish Federation of Women Social Democrats was formed at the beginning of the twentieth century with the intention of support- ing but also influencing the politics of the Social Democratic Party. The Federation advocated the same antimilitarist and pacifist standpoint as the party but was always seen as the most pacifist of the party's affiliations.[1] Swedish women gained the voting rights in 1921, and the Federation grew rapidly during the 1930s, when the Social Democratic Party ran the Swedish government. The Social Democratic women became an important voice in Swedish politics. They debated the declining birthrate, work prohibitions for married women, skilled and professional women's positions in society, and women's possibilities to be elected as members of parliament. In the 1930s, these issues were discussed in the global context of increasingly unstable international relations and the imminent threat of war. The potential for conflict ran over into domestic policy, forcing the discussion of such issues as the growing militarization of society in preparation for a possible future con- flict, including civil defense, gas masks, and air-raid precautions. The Federation believed in the need to defend their country but also to defend fundamental political standpoints such as neutrality, international coopera- tion, antimilitarism, and pacifism. The juxtapositions between the needs for defense and the pacifist standpoints of the women offered very real political and personal dilemmas, particularly when the Second World War drew in Sweden's nearest neighbors and threatened the nation's avowed neutrality.

According to the April 1939 issue of *Morgonbris* (Morning Breeze), the publishing organ of the Swedish Federation of Women Social Democrats, "the majority of women" had long been pacifists. However, *Morgonbris* simultaneously asserted that the potential for Sweden and Europe as a whole to be plunged into a state of "total war" challenged this perception. As a result, despite living in a neutral country, Swedish women and children, who had previously felt themselves safe, must prepare themselves for the potential dangers of becoming immediate targets of war, particularly of aerial warfare. If Sweden went to war, the technological and ideological developments that characterized modern warfare would inevitably draw women (and their families) in, whether they were pacifists or not. It was, therefore, important that they familiarized themselves with the civil defense system, such as using shelters and gas masks, helping out with first aid, and implementing air-raid precautions.[2] "Preparation for the worst" was the message that the Federation communicated to its readers on the eve of war.

A year later, when all Sweden's neighbors were either occupied or at war, *Morgonbris* published an advertisement encouraging women and their families to subscribe to defense loans. The advertisement's text appropriated the same references—to the past and to peace—as the Federation's prewar publications that exhorted total pacifism. Before 1939, peace and the establishment of good relations between peoples had been the primary goal of the Women Social Democrats who opposed military expenditure of any kind. However, when Sweden's neighbors began to be threatened, attacked, and occupied by foreign powers, these pacifist ideals were seconded to ensuring Sweden's self-defense, so that freedom and democracy could be saved for the future. Improving social conditions within Sweden became, in the eyes of the Federation, just as irresponsible and frivolous in wartime as redecorating your home or buying new furniture.[3]

The aim of this chapter is to understand how this major attitudinal shift from staunch antimilitarism to active engagement in the civil defense came about. On the eve of war in 1939, the Federation of Women Social Democrats, with its 648 branches and 26,307 members, faced numerous challenges in considering the mobilization of the Swedish population for war.[4] The chapter deals with the Federation's actions in response to the invasion of Sweden's neighbors by Germany and the USSR from September 1939 onwards. It analyzes the changing attitude of the antimilitarist Federation to civil defense during the first year of mobilization (1939 to 1940). It also looks at the ways in which the women Social Democrats promoted national and civil defense initiatives (such as coordinating air-raid drills and knitting socks and rifle mittens for National Guard soldiers) from their ambivalent positions toward defense and defense propaganda.

Furthermore, the chapter focuses on the challenges facing the Federation in finding a place for "traditional" women's wartime tasks and the demands of modern professional and trained women within women's public emergency work. All these issues highlight, above all, that even in a nation removed from war, as neutral Sweden ostensibly was before and during the Second World War, war concerns, defense issues, and other military matters were a pressing concern to Sweden's citizenry.

Antimilitarism and Working for Peace in the 1920s and 1930s

By 1939, Social Democratic women in Sweden had a long tradition of working for peace, and they had chosen various allies in their work. In addition to being loyal to the antimilitaristic attitude of their own party, they cooperated with socialist women in other countries, especially for International Women's Day (held annually around 8 March since the early 1900s). They also joined together in broad coalitions for peace with liberal women in Sweden.[5] Knowledge of the Swedish Federation of Women Social Democrats' cooperation, pacifist activities, and attitudes in the years since their foundation is crucial to the broader understanding of some of the problems that emerged with regard to Sweden's mobilization for the Second World War.

As early as the First World War, Social Democratic women played key roles in promoting peace and pacifism. Prominent Social Democratic women, such as Anna Lindhagen, Signe Svensson (Vessman), Anna Sterky, and Agda Östlund, took part in the "Women's Peace Sunday" action in 1915, an antiwar campaign that drew in women from the entire political spectrum, including liberals and conservatives. The action gathered 88,000 women who, in 343 different places throughout the country, agreed on a demand for neutral mediation between the warring parties. The address that was read out at all 343 peace meetings that Sunday noted that it would be unnatural for civilized countries to "periodically enter a state of war with intervals of peace that would be used for rearmament." Instead, a new order must be established where "representatives of people of all countries" are provided with the opportunity to control foreign policy and disputes between countries are decided through arbitration—not fighting.[6] Such actions were not extraordinary during the First World War; women in neutral countries throughout Europe clamored for peace and undertook all manner of initiatives to bring about an early end to the war and suffering. For the Federation of Women Social Democrats in Sweden, particularly, it sparked a long-term focus on pacifism and peace issues.

At the end of 1918, in conjunction with President Wilson's attendance at the peace negotiations in Europe, the Central Co-operation Committee of the women's peace organizations collected names in support of Wilson's peace program. The Federation of Women Social Democrats was one of many women's groups represented in the Central Committee, alongside the Swedish Section of the International Women's Committee for Lasting Peace, the peace division of the White Ribbon Society, the Swedish women teachers' peace group, the committees of the National Association for Women's Political Suffrage, and the Liberal Women's Organization. A large public meeting was held in January 1919, and nearly 50,000 women's names were collected from all around the country in support of a just peace built on legal legislation, arbitration between nations, and the abolition of militarism.[7] In the immediate aftermath of the First World War and despite Sweden's neutrality in that war, Sweden's women called for universal peace. The Federation of Women Social Democrats, despite their radicalism in other areas, upheld a conventional message in this respect. Furthermore, much of the Federation's activities with regard to peace and antimilitarism remained within the shadow of the wider Social Democratic Party's initiatives. Nevertheless, it did so from a particular women's perspective and within the network of women's peace movements that had been so vocal during the First World War.

During the election campaign of 1924, the defense issue was a major preoccupation in Swedish politics. Importantly, this was the first time since the introduction of women's suffrage in 1921 that women had the opportunity to participate in an election where defense issues were debated. Within the context of this charged atmosphere, the Federation rallied around the slogan "No More War" and organized another Peace Sunday. The August 1921 issue of *Morgonbris* was dedicated entirely to the message of peace and reported not only that a peace demonstration would be held but also that the executive committee of the Social Democratic Party had designed a badge with a palm leaf for peace. It also recommended that a joint action group against militarism be formed.[8] In a separate circular, the committee of the Federation of Women Social Democrats appealed to all their members to attend the big peace demonstration without fail. The text stressed that most women had always abhorred war and desired peace and that it was disarmament, in an international attitude of mutual understanding and cooperation, that would pave the way to lasting peace.[9]

While the Federation of Women Social Democrats lent its support to antimilitarism, the Social Democrat prime minister Richard Sandler at the same time encouraged the women to come up with something new in the sphere of peace, now that women were able to use their right to vote and thus were participating in public life. He was of the opinion that they should not resemble the men too closely; instead, they should be independent and make their own statements regarding peace work. He delivered this message during an information week sponsored by the Swedish branch of the Women's International League for Peace and Freedom in January 1925.[10] As mentioned above, women's suffrage in Sweden came about late, and in 1924, when the defense issue was at stake, the women had only had the opportunity to vote once before in 1921. The result of the election of the Swedish Social Democrats in 1924 was a huge reduction in the Swedish defense system. In the following year, the Swedish Women's Left Wing Federation took a more radical pacifist standpoint as a result of Sandler's encouragement to march to the beat of their own drum.

To celebrate International Women's Day in 1930, the committee of the Federation of Women Social Democrats offered their local branches peace flowers to sell with the accompanying tag "Women's unanimous contribution—the assurance of peace" and hoped that within every branch a separate peace group would be formed.[11] During the same year, the committee of the Federation also recommended that local branches order a course of study on peace issues from the Workers' Educational Association, which included the following subjects: "the necessity for peace work," "the causes of war," and "the elimination of war through peace work."[12]

Although the study of peace was crucial to the Federation of Women Social Democrats, it was already apparent by 1934 that issues of war, militarism, and peace had to be handled with care. For example, when "The International" in 1934 appealed to all its affiliates to protest against the rise in armaments, the Federation adopted a "wait-and-see" policy. In their discussion following the appeal, Social Democratic women came to more fully comprehend the changes in the international situation that necessitated an in-depth discussion of the "military issue" in the immediate future. Up until 1934, the Federation had been interested in peace work, both in national and international arenas, but mostly in the shadow of the party. In the mid- to late 1930s, however, the situation in the world changed, the Federation grew rapidly, and took more independent steps.

From Peace Activism to Civil Defense Protests

The need for a civil defense strategy in Sweden was not investigated until 1936, and a law regulating civil defense did not appear until a year later. In 1934, while the appeal for international cooperation on peace was unfolding, Swedish authorities had not yet taken any measures concerning the vulnerability of the civilian population in the eventuality of aerial warfare. The issue of the necessity of civil defense and what it entailed remained unsettled and, thus, was open to debate. Initially, it was only the Red Cross who printed a pamphlet telling civilians how to protect themselves in case of attack.[13]

During the early 1930s representatives of the various women's peace movements—but not the Federation itself—carried on a fierce dialogue with representatives of the army as to whether it was actually possible to build up a sufficiently satisfactory civil defense. The Swedish Women Left Wing Federation and the Swedish branch of the Women's International League for Peace and Freedom, for example, argued that gas masks were few and expensive and that air-raid shelters offered insufficient protection against chemical weapons. Because women made up at least half of the civilian community that was thenceforth to become involved in the preparations for war, these two organizations believed that it was time for women to take a stand.[14] While it was not immediately involved in these discussions, the Federation of Women Social Democrats was conscious of the debate and acknowledged that there was a need to discuss the role of civil defense in Swedish society. When, in 1935, they launched the campaign "Women's outlook on society—The future of humanity," they used a poster that depicted a group of schoolchildren wearing gas masks. The text pointed out that Social Democratic women in all countries had a vital role in bringing about world peace and general disarmament:

If women's outlook on society had been strongly directed against the war, would the future of the human race look as it does now? If Social Democracy had been strongly established among women in all countries, would the future of the human race look as it does now? No! All countries would have had governments that would have been better versed in peace work and the representatives in Geneva would be concentrating on disarmament in all countries instead of on rearmament and the spreading of new methods of warfare that are devastating to humans.[15]

The Federation of Women Social Democrats was, therefore, keenly aware of the function of war and militarism in the Western world. Its members were, above all, antimilitary and propeace. In the context of the rising international tensions of the mid-1930s, however, it was one thing to acknowledge the barbarity of war and the necessity to abolish the production of armaments, and it was quite another to ignore the potential realities of future war for Sweden and Swedish women.

Where is Our *Lysistrata?*

The Federation of Women Social Democrats celebrated thirty years of publishing the newspaper *Morgonbris* in 1934 with events all over the country and took the opportunity within these festivities to make a statement against war and militarism. In Stockholm, Aristophanes' play *Lysistrata* formed an important part of the celebrations.[16] At one showing of the play, Federation members occupied the entire Dramaten, the Swedish National Theater. The program began with a separate prologue about the Federation's battle against war and violence and their fight for freedom and democracy.[17] Subsequently, three women discussed their reflections on the performance in *Morgonbris* under the heading of "Where is Our Lysistrata? Is a Modern Repetition of Lysistrata Conceivable?" The poet Karin Boye answered the question by saying that it was a myth that there was some form of a "woman's special peacefulness" that crossed class boundaries and stood women apart from supposedly "militarized" men. However, since the sexes played different roles in the mobilization for warfare, her opinion was that women should, as far as possible, show solidarity with one another in opposing all war. They could find the strength to achieve this in embracing the myth and symbol of *Lysistrata*. Another contributor, the party secretary Hulda Flood, however, was skeptical about the possibility of a joint peace action among women. She opined that during the First World War women had not been asked their opinions about war, but since then many nations had granted women the right to vote. The impact of women's votes was bound to be diluted, however, by the continuing centrality of men in positions of power throughout the world and the fact that many women relied on men's political decisions while occupying themselves with the trivial tasks of decorating their homes. However, Flood did think that perhaps concern over the situation in the world today could bring women together and make them conscious of the dangers of war and militarism. Dr. Andrea Andréen, a member of the Swedish Women Left Wing Federation and a Social Democrat, held a more positive view. She considered that men were more taken with the simple weapons of *Lysistrata*'s times than with those of modern war technology and that modern technology thus was on the side of modern women. In addition, women were now in a position where they had technical assistance that gave them the power of refusing to bear children at times when violence was prevalent.[18] The message of the drama *Lysistrata,* women refusing to make love to men as a protest against war, could in a Swedish context be seen as a political statement and as an explanation of the low birthrate. However, it could also be interpreted as an invitation to women to act against war. That *Morgonbris* published three such varying opinions of the play and of the role women should take in countering war illustrates the array of opinions prevalent within the Federation. That there was no singular unified standpoint among individual members heightened the ambivalence of the Federation with regard to taking a clear stand on issues of vital importance to Sweden's national defense.

Still, it did not stop women Social Democrats from protesting the existence of war and militarism in the name of *Lysistrata*, which they did by supporting the collective action entitled "Women's unarmed revolt against war" during the summer of 1935.[19] Essentially, the gathering was a protest against both the rearmament of Sweden and plans for building up civil defense. The promoters did not consider that it would be possible to protect civilians in the event of chemical warfare. Therefore, if women unanimously refused to use gas masks and refused to go down into shelters in the event of air attacks, then men would become aware of their responsibility, lay down their weapons, and return to the negotiating tables. War could, thereby, be prevented.

Since war was not imminent in 1935 (at least from Sweden's point of view), the Swedish Women's Left Wing Federation and the Swedish branch of the Women's International League for Peace and Freedom organized an election for a representative women's assembly that would prepare a resolution to be sent to the League of Nations. Approximately 20,000 women took part in this process, and the elected assembly duly demanded of the League of Nations that the adjourned Disarmament Conference should be resummoned and that the League should do everything to secure world peace. A delegation traveled to Geneva to present the resolution to delegates at the next general meeting of the League of Nations and to try to start an international revolt among women against war. The election was not an initiative of the Federation of Women Social Democrats, although Signe Vessman, the president of the Federation, and Kaj Andersson, the editor of *Morgonbris,* did participate. Other prominent Social Democratic participants included Hulda Flood, Signe Höjer, Ulla Alm [Lindström], Alva Myrdal, Herta Wirén, and Disa Västberg.[20]

Such protests achieved little, however. In 1936 the government initiated an investigation into civil air defense. The National Air Defense Association and the first voluntary civil air defense associations were also established in Sweden at that time. Laws concerning civil air defense were passed in 1937, and in 1938, the right of disposition law was passed, which allowed municipal authorities to commandeer personnel for civil defense duty in time of need.[21] Despite the initial actions against civil defense taken by women's groups, including the Federation, by the late 1930s protests about the appropriateness of civil defense declined. Only a very few women, mostly from the radical Swedish Women's Left Wing Federation, continued to discuss the possibility of individual acts of defiance and protest, such as refusing to dim their lights or covering their windows with dark curtains during air raids.[22] It seems that even the more resolute of antimilitary women had quickly (if not quietly) become accustomed to the fact that if war came, no amount of passive or active protest by them would actually stop the bombs from falling.

Our Purpose Is Not to Become Cannon-Firing Women

Although the Federation of Women Social Democrats joined the protest "Women's revolt against war" and took the most antimilitaristic stand in the Social Democratic Party, its members, on the whole, did not identify themselves as radical pacifists. Historically, they had, however, viewed humanitarian groups, such as the International Society of the Red Cross, as defense organizations focused on war rather than on peace. As a result, membership of the Federation all but precluded simultaneous membership of the Red Cross. However, with the mobilization of Sweden and the Soviet attack on Finland in 1939, the Federation changed its official position toward defense organizations such as the Red Cross and the Sweden Women's Army Auxiliary Corps.

The issue was not a simple one, but it was also not unprecedented. During the First World War and again in connection with a shooting in Ådalen in 1931, Swedish women of antimilitarist and pacifist principles had been forced to reexamine their position on civil defense and their refusal to support the bearing of arms. In 1915, when two members of the party from Malmberget in the north of Sweden had joined the women's association of the Red Cross, the affected local branch placed the matter before the central committee in Stockholm, who discussed whether membership in the Red Cross constituted grounds for exclusion. The Red Cross was seen as an organization that worked to reduce the effects of war, and its activities were not interpreted as peace work. The committee decided that it was each individual woman's private concern to determine whether she had the "time and energy" to give to two organizations and that the conditions of war that prevailed "demanded such latitude towards groups such as the Red Cross."[23]

In 1931 in Ådalen, five unarmed unemployed men were killed in a demonstration when soldiers were commandeered to keep order among unarmed demonstrators.[24] Arguments raged among party members as to what right-wing interests were being served in controlling the workers by using the army. *Morgonbris* contributed to the subsequent debate in its publication of pictures depicting Women's Army Auxiliary Corps shooting rifles. The photographs were captioned with the statement: "Right-wing women learn to handle weapons."[25] These pictures were a part of a broader development of the women's voluntary defense service, which was not satisfied with creating comfort and well-being for male soldiers but encouraged women to openly participate in overtly militarist tasks.

Given the Federation of Women Social Democrats' openly antimilitary stance throughout the 1920s and 1930s, when the issue of civil defense against air raids and chemical warfare became topical in the late 1930s, different branches across the country called upon the central committee to establish a coherent policy. The Swedish government, of course, argued that everyone should be involved in air-raid precautions, because everyone could be a target. But Federation members were concerned with reconciling the stand taken by them during the "Women's unarmed revolt against war" in 1935 with the strong possibility that their nation could be under attack and their own lives could be at stake. It seemed counterintuitive for them to refuse to support civil defense when that might be the one thing that could alleviate the suffering of a possible war situation. In other words, many Social Democratic women in local branches were revisiting their stand on war and wished to become guardians of neutrality in line with their Social Democratic government's requests.

After the air-raid precaution law was passed in 1938, the central committee sent out a circular to the rest of the Federation in which it acknowledged the need to alter its previously passive attitude toward air-raid precautions. In the face of possible aggression by fascist nations, so the committee argued, the burden for democratic and nonaggressive countries was the adoption of civil defense measures. This could be construed as being contrary to the pacifist ideals of the Federation. However, given the very real threat foreign belligerents posed to civilian populations, it was acceptable for an individual to both be a member of the Federation of Women Social Democrats and, in a "conscientious and discerning" manner, participate in air-raid precaution activities.[26] But the Federation carefully distinguished between emergency help and defense propaganda:

> The air-raid precaution activities that our members should participate in can be compared with fire-fighting and first aid duties and should not be associated or confused with militia duties or defense propaganda as other parties pursue it. Tasks that can be allocated to women during air-raid precautions scenarios are primarily fire-watching and fire-fighting duties, care of the sick, transport of the injured, helping with the evacuation of children and the elderly, and their care in other places.[27]

On the other hand, the Federation emphasized the fact that it was also of great importance that women, with their "predisposition to peace," ensured "that these new air-raid precaution organizations did not come solely under non–Social Democratic leadership and influence, but that a genuine popular spirit prevailed." On a purely pragmatic level, it was believed that Social Democratic women could utilize their participation in civil defense organizations to meet "the unenlightened class of people in society," which would provide new opportunities to spread "enlightenment" and the indisputable facts of the dangers of possible future war with Germany.[28] In other words, Social Democratic women were encouraged to combine an involvement in air-raid precaution activities with their political work.

That air-raid precautions work should not be allowed to come under the control of non–Social Democrats was taken up again by the committee of the Federation of Women Social Democrats in 1939, when individual members were encouraged to take an active part in civil defense. However, neither the Federation nor the local branches could participate as groups, since civil defense was perceived as a voluntary defense task in which individual women participated on the basis of their personal convictions.[29] The message the Federation gave to the individual members was that the growing militarization of the world and the increase of international violence had forced the Social Democratic women to be a part of the Swedish defense system. The Federation wrote:

> Our purpose is not to become cannon-firing women or anything of that sort, on the contrary, anarchy, the threat of violence, armament, etc. in the world, emphasize more strongly than ever the need for work for arbitration, co-operation, disarmament and peace; however, before we attain this, we must protect our neutrality and freedom from the powers of violence.[30]

Allowing its members to volunteer for air-raid precaution work in 1938 opened the floodgates for other forms of active participation in civil defense, and by late 1939, the Federation encouraged its women to get involved in all manner of essential civil defense work. The Federation also joined the Women's Emergency Committee, a comprehensive coalition of women's organizations that had been established the previous year. After the outbreak of the Second World War, the Federation made it clear that emergency work among women was essential on humanitarian grounds, for example, in replacing mobilized men as agricultural laborers to safeguard the nation's food stocks or in preparing for the evacuation of cities. The Federation trod a fine line in finding and maintaining the difference between work for the civilians as humanitarian work and work as preparations for war. They were forced in this way to straddle their activities between the civilian sphere and the military sphere, which involved finding a balance between what they urged the women to do as members of the Federation and what they thought the women could do as individuals. Women of the Federation needed to construct new ways of regarding peace work within the frame of total war. They were going to have to be involved in a total war, but how could they then continue to look upon themselves as a federation advocating peace? The experiences of the Spanish Civil and Russo-Finnish Wars became exemplars in terms of the necessity to involve women in civil defense and, according to the prevailing gender pattern, in the humanitarian care of children, the sick, and the elderly. The Federation told its branches that they should organize appropriate courses with the help of a correspondence school, Red Cross members, nurses, midwives, and scout leaders.[31]

In addition to work on air-raid preparations, food production, and evacuation plans, in the New Year of 1940, the Federation sent out the following appeal: "Form groups to collect warm clothing for our guardians of neutrality." In this project, it seems that women, both on an individual and on a branch basis, could participate in the country's neutral war effort without any conflicts of interests. Protecting neutrality was very much aimed at future peace rather than fuel for the military machine despite the fact that it was the army that would be the beneficiaries of such activities. *Morgonbris* filled an entire page with knitting instructions titled "Patterns from the Guardians of Neutrality" for kneecaps, gloves, scarves, wrist-warmers, and rifle mittens "with a thumb, a forefinger and broad covering for the other 3 fingers . . . [e]asy-to-knit pattern with seams along the sides. Knitted on 2 needles. Use needles no 2 and 3 . . . Cast on 56 stitches ."[32] Branches all across the country were quick to comply. The branch of the town Oxelösund reported that a circle that used to meet to sew and read had been transformed so that the members now knitted while reading. From Vadstena, the members reported that interest in providing for the "guardians of neutrality" was great and that they could acquire yarn themselves. In Emmaboda, they unpicked their cardigans and other old clothes so that it need not cost the members anything to participate.[33]

Despite these clear moves to aid the efforts to protect and defend Sweden's neutrality, there remained unsatisfactory juxtapositions and ambivalences in the Federation's position on what was deemed a more "active" involvement by its members in defense activities, such as in the Red Cross and the Women's Army Auxiliary Corps. In 1939, the local branches of the Federation once more applied pressure in order to get advice about what attitudes they were to adopt regarding "those women's organizations working for our defense." Knitting for Sweden's "neutral guardians" and their weaponry was one thing, personally undertaking war activities was quite another. As it had in 1938, the Federation reiterated that it was up to the individual woman.[34] Still, in 1939, Disa Västberg, the president of the Federation, paved the way for a more wholesale reconciliation of all types of civil defense activity within the Federation's worldview when she thanked the Women's Army Auxiliary Corps for their personal contribution in Finland in the magazine *Idun:*

> That which I have previously never been able to take in earnest has become a dire necessity. I wish to protect my country, I thank the Women's Army Auxiliary Corps, and I value our co-operation across the battle lines.[35]

Västberg's honoring of the Women's Army Auxiliary Corps in 1940 marked a clear break from the past for the Federation of Women Social Democrats. Once Sweden was surrounded by warring and occupied countries (by the spring of that year), defense work could no longer be viewed as an individual task; it concerned the Federation as a whole as it did the nation as a whole.[36] The transition in attitude from being a neutral in peacetime to a neutral in wartime and whether or not to take part in the civil defense of the nation was transformed from a rhetorical argument to a practical necessity. The luxuries of arguing about not taking part in civil defense in a nonthreatening environment were not possible when the reality of war loomed large over Social Democratic women's lives. The transition was made blatantly clear to members when the Federation issued a circular in May 1940 supporting the decision by the Civil Defense Authority to recruit volunteers, both women and young people, for aircraft-warning duties.[37]

Accepting that their members could volunteer for "military duty" (albeit in spotting aircraft and not shooting at them) marked a huge shift for an organization dedicated to the principles of antimilitarism and pacifism, but it was a shift whose implications, for women, gender roles, and the Federation's political beliefs, were left unspoken and unexplored. Aircraft-warning duties were part of the military system and those who volunteered were paid on the same basis as other members of the nation's military forces. That any women Social Democrats who volunteered for this duty would become part of the national defense force was never mentioned in the circular or subsequently.[38]

As the war situation within and outside Sweden became more dire and the possibility of attack grew, the blurring of the Federation's pacifist ideals and policies deepened as well. In June 1940, for example, *Morgonbris* published an appeal encouraging women to sign defense loans in the following terms:

> Many of us considered earlier military budgets and defense costs as something rather unnecessary and negative. If we must appropriate something for military purposes, then we demand the means for social welfare work in similar proportions. Otherwise, we want peace and the establishment of good

relations between peoples . . . Of course, we still desire this, we have no greater wish, but the recent appalling events, not the least in our neighboring countries, have made it clear to us that more important and more imperative than anything else is that we set our defense in order, that we have the means to protect our country, our freedom, our democracy.[39]

During the autumn of 1940, it was again time to knit, but now the army delivered wool for this purpose in ten-kilogram lots. The socks were to be knitted according to a pattern provided and reinforcing yarn was to be knitted in at the heels and toes. The socks were to be sent within two months of the delivery of the wool.[40] The patterns to protect the guardians of neutrality required military precision over and above any aspirations to socialist zeal or pacifist ideals.

Voluntary Women's Work and Modern Qualified Work

Despite the Federation of Women Social Democrats' active encouragement of their members volunteering for war work, a dilemma about the recognition women received for participating in this work remained unsolved. The Federation was duly concerned about the ways in which Sweden's wartime society was appropriating women's voluntary time and failed to recognize the true value of the contributions women actually made. It did not believe that wider society placed adequate value on the time and effort required for many activities, including knitting socks for neutrality. More significantly, the Federation leadership worried that not enough use was made of skilled and professional women's work in these times of emergency.

When the Social Democrat and vice president of the International Federation of Business and Professional Women Alva Myrdal discussed the issue of women performing public defense work in *Morgonbris* in January 1940, in an article entitled "Peace Service During Wartimes," her disappointment was tangible. It is true that she felt that women should make their contribution and support their country. However, she advocated that rather than women's contributions being framed within the spirit of voluntarism, the work of women, like the work of men, should be treated as work, not charity, and remunerated accordingly. Despite the fact that women were incorporated in society's wartime production, they had not been asked to contribute either in act or in word on those committees "that make decisions about our consumption and the evacuation of our homes." She interpreted this as implying that women were not "seriously counted as citizens."[41] Women with qualifications should not volunteer to do work that just anybody could do. She wrote:

However, none of us profits from imagining that a skilled career woman, who could perhaps be hired to follow the foreign press or to be a supervisor in a textile factory, does more for her country by sacrificing her special knowledge and *voluntarily* performing the same civil defense duties as all of the others.[42]

The case for female exceptionalism in time of war was seriously flawed, so Myrdal argued. By implication, she seemed to suggest that the Federation had indirectly contributed to the undervaluing of women in Swedish wartime society and that something had to be done about this.

Myrdal wanted to point out that educated and skilled women were needed in all positions in public life, particularly at this time when Sweden's national well-being was under threat. *Morgonbris* became the medium through which the issue of increased female participation in the war was loudly voiced. On the eve of war, in 1939, the question of which duties women could take responsibility for during a mobilization from a medical point of view had already been discussed. At the time, the newspaper focused on whether women could tackle heavy jobs or whether they had enough technical expertise to undertake certain tasks.[43] Dr. Andréen, who wanted to encourage women to become more interested in technical trades, wrote:

Our country is completely electrified, but no woman—with the exception of one or two engineers—can run a power plant or repair electrical wires. No woman has ever been taught to repair a leak in water pipelines or drains. Nor are there any women employed in the gasworks—other than as typists—and there are no women fire fighters.[44]

Changes in gender patterns were discussed in connection with the female body and the education of women and positions in public life. Another contribution to *Morgonbris* dealt with the fact that in industry women performed the simpler tasks, while men were responsible for the more qualified tasks. Training, however, could change that situation. Certainly, heavy tasks could be simplified by using means of technology, but since women did not have enough technical knowledge, men ended up performing those tasks anyway.[45] Gulli Pauli, also a doctor, continued the debate. She asserted that there were no occupations that were "unsuitable for women" from the health point of view. On the other hand, she claimed that it was not unusual for women to be overworked since they often did all the housework in addition to having an occupation.[46] From a medical point of view, there were no obstacles for women to accomplish work that had been strongly gender signed.

As well as encouraging women to transgress traditional boundaries in their choice of career, *Morgonbris* pursued Myrdal's line that the authorities should make use of women's experience and qualifications on those committees that dealt with economics and evacuation, two areas from which women had traditionally largely been excluded. When the Social Democrat Signe Höjer, the chairman of the Evacuation Committee of the Women's Federation's Emergency Committee, was appointed as expert adviser to the Civil Defense Inspectorate, *Morgonbris* applauded: "A female expert adviser to the Civil Defense Inspectorate has to work with key issues. Women have long been waiting for a woman to be appointed to the Inspectorate. Questions regarding evacuation and air-raid precautions largely concern women and children."[47] In this position, Höjer became a pioneer. Her role on this committee changed the Social Democratic government's position on the role of women as well. From late 1940 on, the government urged everyone in Sweden to work for neutrality and Sweden's security. With the words "Put aside your diffidence, overcome your prejudice and take responsibility," the Federation of Women Socialists further encouraged its women to not only knit socks but also to demand leading posts within civil defense.[48] While the focus on pacifism may have receded from the forefront of the political aims of the Federation during the war, women's involvement in Sweden's defense efforts highlighted another way the Federation could push to increase the prominence and importance of women in Sweden, issues that were ongoing for a society not only at war but also at work.

Tradition and Modernity: A Modern "Laundry Ambulance" for Finland

It is obvious that during the mobilization of 1939–1940, the previously antimilitaristic Federation of Women Social Democrats adapted to the existing situation and the threat of war. To do everything in their power for their country was a point of honor. They justified their change in position by pointing to their unflagging work for peace in the past and to the fact that fascist countries had forced them to become involved in air-raid precautions and civil defense. In the end, it was the war and its impact on Sweden's neighbors that sparked a major change in the Federation's position on women's involvement in defense organizations. While the Federation officially condoned individual women's choices in involving themselves in defense, the organization itself took a more problematic and ambivalent stance. In effect, while the Federation remained the most antimilitaristic affiliation of Sweden's Social Democratic Party, it lay "their essential pacifism" aside during the Second World War, as historian Karl Molin also argues. Their continued pursuit of two key issues helped to maintain the integrity of the pacifist ideals despite the compromises they were forced to make because of the exigencies of war. Looking to the future, the Federation continued to lobby for the prohibition of national service training in schools and argued for a time limit to be imposed on the civil defense law. In this way, the organization followed its own path and held true to the ideal of a world without war and militarism.[49]

The adaptation of the principles of the Federation can also be interpreted as an attempt to reconcile the untenable situation of antimilitarism and pacifism in a time of total war and as an attempt to find a balance between tradition and modernity concerning gender. All over the branches, women willingly pulled out their knitting needles for the neutrality guard and, in the Federation's circulars, the importance of traditional women's areas of work were emphasized, such as responsibility for food and home as well as care of children, the sick, and the elderly. In much the same way that the long tradition of women involved in peace work was invoked through *Lysistrata*, the specificity of women's experience of family, home, society, and war throughout the ages was utilized to defend their antiwar position. However, even in this type of work, there were signs of modernity. The Swedish Federation of Women Social Democrats collected money for the world's first "laundry ambulance," which was donated to the women of Finland. This was a remodeled ambulance refitted with equipment so as to function as a mobile laundry, where soldiers could have their clothes washed by the women operating it.[50] That women were expected to acquire technical experience and assume duties of leadership and responsibility within civil defense can also be interpreted as a trend toward modernity.[51] This was a demand that the Women Social Democrats as well as the non–Social Democratic women's associations in the Women's Emergency Committee were committed to.[52] Regardless of whether at peace or at war, the position of women in society needed to be evaluated, understood, improved, and valued.

In conclusion, it is possible to interpret the actions of the Federation of Women Social Democrats' during Sweden's mobilization leading up to the Second World War as an adaptation to the politics of the Social Democratic Party concerning neutrality and the security of the country. These actions can also be seen as the formation of their own negotiation between the promotion of traditional gender tasks and lobbying for the recognition of the value of women in modern roles of leadership and cooperation in times of emergency. The importance of my study, however, is to shed light on how concepts of neutrality and gender worked together within the Federation of Women Social Democrats in Sweden in the 1930s. On the one hand, it shows the ways in which Social Democratic women were able to work to help the civilian population and to protect the country, and on the other, it highlights how fundamental the concept of neutrality was to this group of Social Democratic women who remained noncombatant and did not want to have anything to do with the defense system but were forced to be involved. From a neutral gender position, they could criticize the war and the suffering it caused both within and outside Sweden.

Despite their involvement in the active defense of Sweden's neutrality and the alleviation of the stresses of war outside the country, the Federation of Women Social Democrats remained a vital voice advocating peace. The organization did not abandon its pacifism per se; it merely adapted it to fit within the trying circumstances of "total war." When the Federation organized an awareness-raising meeting "Against total war—for peace and reconciliation between peoples" in February 1940 as a protest against the war in Finland, 24 different women's groups assembled in the Concert Hall in Stockholm.[53] The message of peace remained the driving force behind the meeting as it did behind all the Federation's activities during this war and in the Cold War years to come.[54]

Notes

1. Karl Molin, Försvaret, *folkhemmet och socialdemokratin. Socialdemokratisk riksdagspolitik 1939–1945.* Stockholm, 1974, pp. 56–59, 62, 81–83.
2. "Vad var och en bör veta," *Morgonbris.* 4, April 1939.
3. Birgit Magnusdotter Hedström, "För övrigt ville vi fred och folkförbrödring," *Morgonbris.* 6, 1940, p. 18.
4. Hulda Flood, *Den socialdemokratiska kvinnorörelsen i Sverige.* Stockholm, 1960, p. 322.
5. For more, see Irene Andersson, *Kvinnor mot krig. Aktioner och nätverk för fred 1914–1940.* Lund, Sweden, 2001.
6. *Föredrag vid de svenska kvinnornas fredssöndag.* Flory Gate's collection, vol. 8, Kvinnohistoriska samlingarna, KS, Göteborgs Universitetsbibliotek (GUB).
7. Resolution, International Kvinnoförbundet för Fred och Frihets samling, L IIc:1, KS, GUB.
8. Agda Östlund, "Fred," *Morgonbris.* 8, August 1924, p. 2.
9. Letters to the local sections from SSKF's board of directors, 9 July 1924, outgoing circular with appendices, 1908–1930, B II:1, Arbetarrörelsens arkiv (ARAB).
10. Report from IKFF's Peace Week, *Morgonbris.* 2, February 1925.
11. "Meddelande för Internationella Soc. Dem. Kvinnodagens hållande," circular no. 1, 1930, SSKF, outgoing circular with appendix, 1908–1930, B II:1, ARAB. It is unclear whether or not these moves were undertaken. The White Ribbon organization formed a Peace Department, and it is possible that they also had local groups. However, it is unclear whether or not local chapters of

the Women Social Democrats undertook what their committee was urging. This issue is clouded through a lack of investigation into Swedish women's history before the 1930s.

12. "Fredsfrågan såsom studieämne i cirklarna," circular no. 6, 1930, outgoing circular with appendix, 1908–1933, B II:1, ARAB.

13. Svenska Röda Korset, *Civilbefolkningens skydd mot gasanfall från luften.* Stockholm, 1929 (1935).

14. "Inför svensk upprustning," *Tidevarvet.* 20, 1934, pp. 4, 10; Naima Sahlbom, *Hur skall det gå med civilbefolkningen? Gasmasker och bombflyg.* 1934.

15. Accompanying appendix to a circular, SSKF, styrelse- och vuprotokoll 1934–35, A2:4, ARAB. All translations in this article are my own.

16. "Morgonbris jubileum firas med 400 fester," *Morgonbris.* 11, November 1934. Hjamar Gullberg and Ivar Harrie, *Aristophanes Lysistrata: A Women's Drama for the Swedish Theatre.* Stockholm, 1932.

17. Hulda Flood, "Var är vår Lysistrate? Kan man tänka sig en modern upprepning av Lysistrate?" *Morgonbris.* 12, Christmas 1934, pp. 7–9; Hulda Flood, *Den socialdemokratiska kvinnorörelsen i Sverige.* Stockholm, 1939, pp. 53–54.

18. Flood, "Var är" pp. 7, 10–11.

19. See: Andersson, *Kvinnor mor krig. Aktioner och nätverk för fred 1914–1940.*

20. Ibid., p. 237.

21. Ibid., pp. 245–247.

22. Ibid., pp. 267–274.

23. Protokoll 24/5 1915, SSKF, styrelse-och vuprotokoll 1906–1935, A2:1A, ARAB.

24. Roger Johansson, *Kampen om historien. Ådalen 1931. Sociala konflikter, historiemedvetande och historiebruk 1931–2000.* Stockholm, 2001.

25. *Morgonbris.* 10, October 1931, pp. 16–17.

26. Ibid.

27. Circular no. 7, April 1938. SSKF, outgoing circular with appendices 1934–1939, B I:2, ARAB.

28. Ibid.

29. Circular no. 8, 22/3 1939. SSKF, outgoing circular with appendices 1934–1939, B I:2, ARAB.

30. Circular no. 10, 22/8 1939. SSKF, outgoing circular with appendices 1934–1939, B I:2, ARAB.

31. Circular no. 1, 15/1 1940. SSKF, outgoing circular with appendices 1940–1947, B II:3, ARAB.

32. *Morgonbris.* 2, February 1940, p. 13.

33. *Morgonbris.* 2, February 1940.

34. Circular no. 8, March 1940. SSKF, outgoing circular with appendices 1940–1947, B II:3, ARAB.

35. *Idun.* 52, 1939, p. 4. Positive statements about the Red Cross and the Women's Voluntary Defense Service are also found in: *Morgonbris.* 10, 1939.

36. *Idun.* 52, 1939, p. 4.

37. Circular no. 11. SSKF, outgoing circular with appendices 1940–1947, B II:3, ARAB.
38. Ibid.
39. Birgit Magnusdotter Hedström, *Morgonbris*. 6, 1940, p. 18.
40. Communication 15/10 1940. SSKF, outgoing circular with appendices 1940–1947, B II:3, ARAB.
41. *Morgonbris*. 1, 1940, pp. 4–5.
42. Alva Myrdal, "Fredstjänst i krigstid," *Morgonbris*. 1, 1940, pp. 4–5.
43. "Var kunna kvinnorna ersätta männen?," *Morgonbris*. 4, 1939.
44. "Vad kvinnor borde få göra," *Morgonbris*. 4, 1939, p. 16.
45. "Böra kvinnor få syssla med sådant?," *Morgonbris*. 5, 1939.
46. Gulli Pauli, "Finns det yrken, som äro olämpliga för kvinnor?," *Morgonbris*. 6, 1939.
47. *Morgonbris*. 1, 1940, p 17.
48. Maj Jarke, "Kvinnlig beredskap, kvinnlig sakkunskap,'" *Morgonbris*. 4, 1940.
49. Karl Molin, *Försvaret, folkhemmet och socialdemokratin. Socialdemokratisk riksdagspolitik 1939–1945*. Stockholm, 1974, pp. 56–59, 62, 81–83.
50. "Världens första tvättambulans överlämnad till de finska kvinnorna," *Morgonbris*. 4, 1940.
51. The discussion can also be said to be a continuation of the discussions of the Women's Labour Committee at the end of the 1930s: Renée Frangeur, *Yrkeskvinna eller tjänarinna? Striden om yrkesrätten för gifta kvinnor i mellankrigstidens Sverige*. Lund, Sweden, 1998.
52. For example, Ruth Stjernstedt, *Kvinnor och försvar. Medborgarkunskap om riksförsvaret 22, Riksförbundet för Sveriges Försvar*. Stockholm, 1945, pp. 54–55.
53. *Mot det totala kriget för fred och folkförsoning. Anföranden hållna vid kvinnoorganisationernas stora opinionsmöte i Stockholms konserthus den 12 februari 1940*. Stockholm, 1940.
54. For example, in the 1950s, Women Social Democrats challenged the Social Democratic Party by refusing to support the development of Swedish nuclear weapons: Gunnel Karlsson, *Från Broderskap till systerskap. Det socialdemokratiska kvinnoförbundets kamp för inflytande och makt i SAP*. Lund, Sweden, 1996.

CHAPTER 3

Bride Ship, Brothel Ship: Conflicting Images of War Brides Arriving in New Zealand in the 1940s

Gabrielle A. Fortune

At the end of the Second World War, banner headlines in the New Zealand and Australian press proclaimed the regular arrival of troopships packed with returning servicemen. Also on board these ships were parties of women and children; the wives, fiancés, and offspring whom New Zealand servicemen had acquired while on active duty overseas.[1] These foreign-born "dependents" of soldiers were transported in their thousands to their new homeland at government expense. Upon their arrival on the wharves of New Zealand's major ports, these "new" New Zealanders were welcomed, scrutinized, and overwhelmed by dignitaries, journalists, and a host of hitherto unmet family and friends.

The war brides and their children were a topic of fascination and gossip even before they reached New Zealand's shores. News about them and their journeys filled newspapers, although not all of it was positive or laudatory. While, on the one hand, the war brides were heralded as wives and mothers and as brand-new "New Zealanders," on the other hand, the expectations of the receiving families and wider New Zealand society were underpinned by suspicion of the wives' domestic abilities and their "foreignness." Particularly fascinating to the gossipmongers was the supposed promiscuity of these women, who had "wheedled" their way into the hearts of "their" boys. Such rumors implied that the women had taken advantage of the young New Zealand soldiers while they were in a "helpless" or "vulnerable" state when serving their country in a war a long way from home. The reception war brides received on arrival in New Zealand was a reflection of the labels given to the ships transporting them. The names —Bride Ships, Stork Ships, Hell Ships, and Brothel Ships—mirrored the multiple roles projected onto the new arrivals.

The incongruous pairing of "war" and "bride" epitomized the conflicting ideas and images about foreign-born brides that circulated in the immediate postwar reconstruction years in New Zealand. The diametrically opposed representations of war brides—as perfect wives, on the one hand, and as flighty, unreliable, hypersexualized women, on the other—reflected the hopes, doubts, and multiplicity of roles incoming women had foisted on them. Images of war brides as, variously, virgins and whores, wives and seductresses infiltrated the public discourse. The existence of such imagery illustrates, above all, how New Zealanders were trying to cope with the personal and societal changes brought about by the end of the war, the return of their soldiers, and the influx of a large number of "strangers" in their midst. For their part, war brides selected, shaped, and discarded these depictions as they saw fit, but it was undoubtedly difficult for them to "achieve unequivocal success" at maintaining their own identities in this judgmental and frequently contradictory environment.[2] War brides were similarly bemused by how their reputations were tainted by the names the press assigned to the ship from which they disembarked. The "bride" and "brothel ship" labels in the title of this chapter are, in this way, indicative of the speculation about the disparity between the brides' wartime experiences and their worthiness to be wives of New Zealand servicemen.

This chapter places the experiences of the women who married New Zealand soldiers within the wider scholarship of war and its diffuse impacts.[3] The problems of subsuming individual identities and personal histories under the label "war brides" parallel, in many ways, the effacing of individuality so important to the functioning of twentieth-century armies. War-bride marriages are poignant illustrations of the social upheaval brought about by the Second World War. This upheaval opened the most personal relationship between a man and a woman and the intimacies of the family up to invasion by the military, media, and state. Most New Zealand servicemen would not have had an opportunity to meet foreign women except for the war.[4] Even for women not directly "at war" or even "near war," it was war service that brought them in contact with their husbands and ultimately with a new society and lifestyle that proved challenging and often alienating. Inevitably, their decision to marry a New Zealander transformed their lives in fundamental ways.

Furthermore, the polarization of gender roles that has been attributed to war mobilization is clearly visible in the war-bride experience. Concerns about the reassimilation of New Zealand women into "normal" peacetime pursuits were amplified by the presence of the newcomers. War brides, with their own histories of war work and war service, generated doubts in a society in flux—their "wifely" qualities and willingness to relinquish wartime roles were questioned. Bridging the gap typified by the New Zealand public's representations of the war brides as the ideal woman versus the femme fatale complicated their transition into their new homes. Even after the initial publicity around their arrival subsided, the fact that their marriages emerged out of a world at war, where normal life for many had been turned upside down, left a legacy associating many war brides with loose morals, prostitution, and a lack of domestic capability.

Still, even though the marriages of the women were born out of the wartime turmoil, their journeys to New Zealand coincided with genuine efforts by the New Zealand government to return social and economic stability to the country and its residents. To this end, besides being expected to fill caring and nurturing roles as wives, mothers, daughters-in-law, and homemakers, New Zealand politicians cast war brides as economic assets serving national and international purposes. In particular, they were described as a tradable commodity in exchanges between Great Britain and New Zealand, as ambassadors between Commonwealth countries, and as contributors to the strengthening of "the Empire." Sir Patrick Duff, the British High Commissioner in New Zealand, said of war brides leaving Britain that

> [t]he whole question [of migration] is rather a delicate topic. So far as Britain is concerned we have a great deal of [reconstruction] work to do and we shall need every pair of able hands. On the other hand, from the point of view of the peace of the world, it is a good thing to have the Dominions and British outposts strengthened by British stock.[5]

Conceptualizing the traffic in war brides in this way paradoxically recognized their migration as part of a worldwide economic and political exchange while simultaneously assigning them to a reproductive role. The migrants boosted population numbers and, to the eyes of the government at least, represented a "good investment." In recognition, the government paid the passage of servicemen's wives, widows, and children. In contrast, servicemen requesting passage for a fiancée had to pay a deposit against the transport costs of their intended bride, which was refundable on marriage.[6] The government's policy vis-à-vis servicemen's fiancées suggests the extent to which marriageable women were reduced to commodity status, blurring the lines between bride and brothel further.

Background

At the end of the Second World War, the Allied governments prioritized the repatriation of troops but in addition had to provide passages for any dependents the troops had acquired while on overseas duty. War brides' migration was an international exchange that resulted in the relocation of hundreds of thousands of women between 1942 and 1952. They moved in great numbers from Britain to the Dominions but also from Britain, Australia, and New Zealand to the United States, from Europe to the United Kingdom, and so on. The magnitude of the movement is illustrated by the fact that over 100,000 women left Japan for the United States; 80,000 British women and 14,000 children also went to the United States; Canada received 40,000 British women and approximately 20,000 children.[7] New Zealand was part of this international exchange. While marriage on active service was officially banned, as many as 4,000 servicemen applied to have spouses and fiancées transported "home." The 4,000 women (and 1,000 children) who entered New Zealand came mainly from Britain and Canada, with smaller numbers from Italy, Greece, Crete, and other European countries, the Middle East, North Africa, and Japan.[8]

New Zealand's War Cabinet dictated the conditions under which marriages of servicemen were permissible, and the Defence Department vetted intending brides (and grooms), restricted opportunities to marry, granted wives' allowances as they saw fit, and controlled access to transport. The government devised policies couched in terms of immigration quotas about "acceptable" and "unacceptable" spouses. In this way, the government exercised powerful economic, legal, and administrative control over the lives of the war brides and shaped gendered perceptions of them as well. Becoming engaged to a New Zealand serviceman transformed them from love interests of individuals into a part of a military operation. The Defence Department had a vested interest in controlling the relationships of servicemen, as the wartime marriages of military personnel were anathema to tacticians focused on victory. Major-General W. G. Stevens (commanding officer in the Middle East) wrote that officers were plagued by the ongoing "very great expenditure of thought and work out of all proportion to the numbers involved and at times such as the [military] crisis of 1942 was definitely embarrassing to our war effort."[9] In the face of ever-changing government guidelines, the Defence Department attempted to accommodate individuals' wishes to marry while keeping its eye firmly fixed on military priorities.

So while most war-bride marriages started out as romantic encounters, the private commitments to marry foreign servicemen proved more complicated than war brides might have imagined, when their marriages were treated as events in the "public" interest. These women unwittingly invited Defence Department and government officials to adjudicate on their personal decisions. Marrying and migrating as a war bride under the auspices of the Defence Department imposed constraints that shaped war brides' experiences. Conditions of entry to New Zealand were dictated to war brides, and politicians framed war-bride migration in ways that fitted postwar reconstructive, economic, and population strategies. The multifaceted projections of war brides and their migration by these agencies suggest how their experiences could be utilized to fit various cultural, social, and political agendas. War brides were never simply choosing love, marriage, and domestic life: they were embarking on a journey that exposed their private affairs to official and media interpretation and intervention.

From the government's perspective at least, gendered expectations of the war brides' roles in New Zealand were made abundantly clear. Information supplied to war brides on board ships, on arrival at the wharfs, and at receptions, both public and private, emphasized their role in New Zealand society as the wives of servicemen and the mothers (and future mothers) of their children. For example, in one *Weekly Review* newsreel of returning fighter pilots escorting their wives and children off the ships with carrycots and other parenting paraphernalia, the commentator remarked that

> proud fathers carry their offspring ashore in the latest portable bassinets. When they left New Zealand they never bargained on this but right now they can't think of a better homecoming. Showing off their bright-eyed-babies means more to them than their deeds in the air.[10]

Returning servicemen were presented as committed to family values to the extent that their war service was almost, but not quite, relegated to second place. War brides in this instance were designated as the supporting cast in the homecoming display of returning war heroes.

Before They Were Brides

Racial and gender stereotypes typified how New Zealanders saw the war brides. While the media played up the romantic whirlwind nature of the relationships between servicemen and their new partners, featuring detailed descriptions of engagements and weddings and of brides who stowed away on ships or were smuggled out of war zones, the characters of the women were always commented on within the framework of the "feminine" qualities they supposedly possessed.[11] Typical of the commendations expressed in this way was that of Reverend Father Leo Spring of Timaru, an army chaplain, returning on the *Tamaroa*.[12] The Italian brides were, he said, "girls [who] would make very good wives. Their whole upbringing had taught them domestic life and how to make a home."[13] While brides of a domestic orientation were welcomed, suspicion of "foreign women" was reinforced by stories that undermined the fairy-tale ideals of white weddings and star-crossed lovers. Promiscuous behavior on board the *Rangitiki* was reported in the press in July 1946:

> A young woman who had been aboard the *Rangitiki*, which arrived yesterday left the ship during its enforced hold-up at Panama and married a US Canal Zone policeman. She was en route to New Zealand to marry a New Zealander who had paid her fare.[14]

News items of this nature, therefore, portrayed some of the incoming women as capricious and unfaithful, bawdy rather than bridelike.

A notable example of these hostilities and of what Ingrid Bauer calls "antithetical pairing" of loyal wives and perfidious war brides can be plainly seen in the following example published in late 1945 and circulated among soldiers attached to the Second New Zealand Division (2NZEF) stationed in Trieste, Italy.[15] In a poem titled "If the Cap Fits," written by a member of the New Zealand Women's Auxiliary Army Corps (WAACs) attached to 2NZEF in Italy and addressed to New Zealand servicemen, Italian women were depicted as femmes fatales or seductresses and were contrasted with the faithful, long-suffering New Zealand girlfriends and wives back home. Soldiers were accused of forgetting their New Zealand girlfriends "Who faithfully are waiting still / Till you come sailing home" in favor of "the girls from Old Trieste" who "glamorise with paint / and go about half-dressed," revealing "their hidden charms."[16] Compounding this view of foreign women as enticing and sexually available, other stanzas emphasized the perceived treachery of Italian women whose motives were considered dubious and whose loyalty was transient. The New Zealand WAACs reminded their male counterparts that "e'er Trieste was won / Before you stormed the town, / Those lasses they were snipers there / Our own lads shooting down."[17] Italian women's paramilitary activities were elided with the succession of sexual partners attributed to them, including Italian soldiers, German soldiers, and finally New Zealand soldiers. The noncombatant WAACs took an active role in castigating the problems of women perceived not only as sexual transgressors but also as transgressors of the gender order in which it is only acceptable for men to be combat soldiers. In addition there were concerns, backed by the New Zealand army authorities, about the reaction of the local population to these liaisons.

Soldiers were discouraged from fraternization by a cautionary tale reminding them that Italian men would not approve of their associating with local women and that a number of Germans "came to an untimely end through trying."[18]

Choosing foreign partners, especially from former enemy populations, threatened to disrupt postwar hopes for harmony and order. As symbols of this potential disruption, war brides became "the other," and they and the men who wanted to marry them were often marginalized as a result.[19] There was a pervasive view that men and women had to be protected from rash decisions, and by extension, New Zealand needed to be protected from the imposition of "undesirables." Suspicions about the trustworthiness of Italians persisted well into the postwar period although they were not the only ones who suffered as prejudices toward nationals of other ethnic and cultural backgrounds were also apparent. Interestingly, British war brides were often intentionally disparaged as "working-class," typified by a mother-in-law repeatedly telling neighbors that her son had married a "bus-conductor's daughter from Birmingham."[20] Japanese war brides were stigmatized as having been "bar girls," and many a British war bride arriving in New Zealand found that smoking alienated her from her mother-in-law in ways reminiscent of how the habit had raised "concerns about women's sexuality" during the First World War.[21] Negative comment could stem from prejudice based on nationality, ethnicity, or socioeconomic background, but in essence any woman associating with servicemen and becoming a war bride could be targeted.

Ship of Brides or Ship of Whores

The contradictory images of war brides as both "brides" and "whores" were reinforced by the descriptions of the ships in which they traveled that appeared in the press. The *Taranaki Herald* announced the docking of a "Bride Ship" in Auckland in April 1946 when the *Athlone Castle* arrived with 700 women aboard.[22] Earlier that year, the birth of a baby on board SS *Rangitata* as it entered Auckland's Waitemata Harbour earned it the label of "Stork Ship."[23] The Melbourne *Age* greeted the arrival of the same ship with the provocative headline "Wives, Wires and Bananas," summing up the excitement on board *Rangitata* as war brides sampled bananas for the first time since the war had begun and queued at the radio office on board to send cables (wires) to their husbands waiting ashore.[24] Two ships that arrived in Melbourne simultaneously en route from Britain were nicknamed the "Mother Ship" and the "Father Ship" by the *Age* because many of the soldier husbands were on the one while their war brides and infant children were on the other.[25] These designations clearly emphasized the married and domestic nature of the war-bride intake and promulgated the notion that their future role in society was as wives and mothers.

Juxtaposed with representations of war brides' domestic and maternal roles were images of them as wayward or complaining women. These less desirable connotations were conveyed by derogatory names assigned to some ships typified by them dubbing the *Marine Falcon* the "SS *Floating Flophouse*."[26] The *Tyndareus*, with an ample complement of children on board, had been called a "Stork Ship" but was also dubbed the "Hell Ship" by the press whose war-bride interviewees reported the lack of adequate facilities on the long journey.[27] The Australian high commissioner E. J. Williams wrote to London that on board the *Stirling Castle*, sleeping arrangements were cramped, there was a lack of quiet spaces for children to rest, shortages of basic medical supplies, severe outbreaks of measles, prickly heat in infants, one death from tubercular meningitis and another from enteritis, from which, he wrote, "nearly all the children were suffering."[28] Pressure for space meant that the swimming pool on the *Athlone Castle* was converted into dormitories and, on the *Dunbar Castle*, returning servicemen slept on tables in the dining room to allow the cabin space to be occupied by war brides and children.[29] The upshot was that cabin space could not be allocated to couples.[30] As far as possible, therefore, married couples were required to travel on separate ships. When they were on the same ship, they were confined to separate decks and separate cabins. Rank brought privilege, allowing officers' wives to socialize with their husbands on board but keeping wives of "other ranks" separated from theirs, sparking complaints of class discrimination. A war bride disembarking from the *Rangitata* in January 1946 said, "I thought we had left snobbery behind when we left England, but it was pretty obvious on this ship."[31]

While fraternization with spouses may have been restricted on board some ships, socializing was inevitable and indeed encouraged on many voyages. Dances, concerts, race meetings, and beauty contests all featured as entertainments that staved off boredom on the six- to eight-week voyage. They also afforded ample opportunity for flirtatious behavior and sexual relationships. Sheila Kirkwood described with some relish the clapping and cheering that ensued when she and Duncan, a returning soldier with whom she spent a lot of time on the journey out, collapsed in a heap on the dance floor when the ship rolled. She also described in detail the sexually charged atmosphere on board when King Neptune and his retinue presided at the "crossing the line [of the Equator]" carnival. In the tropical atmosphere, Neptune's court handed down "sentences" for "misdemeanors" and women dressed in swimwear were doused with flour and sprayed with water, slipping and sliding on greased deck boards to the uproarious applause of servicemen and crew.[32] Episodes and events such as these fuelled gossip and were the staple diet of the onboard magazines that capitalized on intimate or embarrassing incidents to make jokes. Relationships ranged from flirting, casual sex, and intimate shipboard romances for the duration of the voyage through to more commercial arrangements made in organized brothels where women traded sexual favors for cigarettes and alcohol. Unsurprisingly, these activities were reported back to New Zealand, contributing to an image of "bride" ships as "brothel" ships.

Evidence that promiscuous behavior was a reality among some war brides, male passengers, and crew was reported in New Zealand ahead of the *Rangitiki*'s arrival in July 1946 and it became known as the "Brothel Ship."[33] A romantic short story in a troopship magazine published on board the *Stirling Castle* featured an illicit affair between a war bride bound for New Zealand and a married Australian returning serviceman.[34] Examples of fiancées of New Zealand servicemen jumping ship lent an air of credibility to the stories of war brides as undependable and, by extension, undesirable new residents. They were viewed as emotionally fickle and opportunistic—opportunistic in having married a New Zealander in the first instance and fickle in reneging on their commitment in the second. A witness in a divorce case told the court that the war bride "seemed not to care about the place [New Zealand] and was just having a holiday at his [the serviceman's] expense, I think."[35]

Socializing between the sexes on board led to reports of inappropriate behavior and the media capitalized on cases of sexual liaisons, both fictional and real. Thus, while war brides were welcomed as wives, mothers, and homemakers and were expected to conform and assimilate into New Zealand society as seamlessly as possible, they were viewed as foreign and different, even exotic, and often as immoral, misusing the freedom of the long voyage for sexual exploits. "Bride" and "brothel" ships, therefore, coexisted in the popular imagination and complicated the reactions of New Zealanders to the arrival of the brides. War brides were aware of the mixed reaction they evoked, and to deflect bad publicity, they banded together and denied all knowledge of misbehavior by fellow passengers on the journey. Brides of all nationalities arriving in New Zealand closed ranks to protect their joint image, and, later, formed clubs, one of the objects of which was to bolster each other against such sentiments.[36]

From War Bride to Conscientious Housewife

In New Zealand, the gender order had been preserved in spite of the upheavals of war and expectations were that after the war, returned servicemen would quickly reintegrate into family life and that women would resume domestic responsibilities.[37] War brides, therefore, disembarked into a society geared up for the reestablishment of family where the "rhetoric of gender" contributed to the postwar return to normality.[38] In 1944, Frederick Wood, a professor at Victoria University in Wellington, described the postwar ideal as "a straightforward conventional life" in which a woman's role was to be "an excellent and conscientious housewife."[39] Although most war brides had prewar and wartime jobs, they were collectively described as "housewives" by the shipping companies transporting them to New Zealand and as "repatriated dependents" by the Defence Department that paid their fares. Categorizing war brides in these ways suggests that they constituted a group for whom gender acted as "a primary way of signifying relationships of power."[40] The ramifications of these designations as "housewife" and "dependent" placed war brides firmly in the traditional gendered roles they were expected to occupy. As Martha Gardner wrote of war brides migrating to the United States, "domesticity proved the price of admission."[41]

That war brides were destined to fulfill domestic roles as wives and mothers is suggested by the dominant images of them in both government and popular media sources. Yet the wartime careers of these women included service in the British WAAF (Women's Auxiliary Air Force) and Navy, firefighting in the ARP (Air Raid Precautions) in London, transport driving in the ATS (Auxiliary Territorial Service), and for Voina Stewart of Trieste, four years as a dispatch rider in the Yugoslav army.[42] Servicewomen had to take their discharge from their military service before sailing for New Zealand, thus distancing themselves from their war careers and war service and neutralizing any possible "visible challenge to existing gender roles" when they arrived in New Zealand.[43] In stark contrast, their husbands were demobilized only after landing in New Zealand, so that rank and uniform were still significant markers of status on disembarkation.[44] War brides had also been office workers, nurses, dressmakers, shop assistants, factory workers, agricultural workers, interior decorators, and students before and during the war.[45] Nonetheless, a beauty specialist, an RAF (Royal Air Force) nurse aid, an Austrian kindergarten teacher, and a science graduate from Aberdeen University were all listed on ships manifestos entering New Zealand as "British housewives."[46] Two war brides, who between them spoke half a dozen languages and had been educated at the School of Economics in Paris, were also described on shipping lists as "housewives from Egypt."[47]

By the time they arrived in New Zealand, war brides' roles in, and understanding of, war had been subsumed by their new definition as wives—a definition that assisted their entry to a society and a country intent on working to return to "normality." There was, however, a gap between the brides and the nation of which they were to become a part: a liminal space created not only by the physical distance between a woman and her place of origin. This marginality of displacement was exacerbated by a lack of familial support and connections upon arrival in New Zealand, by official policy, and by the reality that did not always match the warmth of the initial public welcome. In-laws who refused to acknowledge the war brides' pasts—especially their war experiences—made it difficult for them to adjust to the social and cultural norms of their new homes. New Zealanders seemed unaware that war brides had surrendered something in coming to New Zealand, abandoning homes, extended families, and careers to do so. Complaints or comparisons made by war brides did not endear them to the local population who considered that they should be grateful to come to New Zealand, which, as they understood it, was so much better than war-torn Europe. In 1947, a New Zealand woman signing herself "Tui" wrote to the *New Zealand Woman's Weekly*: "These Englishwomen [should] have been thankful enough to come to our country and enjoy its privileges and benefits."[48]

Making an Impression

Fashion reports in the media contributed both negative and positive images of the brides. On arrival, fashion, clothes, and appearance were interpreted as indications of war brides' suitability and adaptability, but conservative attitudes quickly became apparent and sent mixed messages to newly arrived brides. In spite of undeniable widespread interest in their overseas clothes, hair styles, and shoes, war brides found they often transgressed local dress codes. Their breaches of local conventions were closely and, sometimes, critically observed. A war bride, who attended a tea party at the Victoria League in Auckland, felt underdressed because all the local women had hats and gloves, which, she said, "we had long since given up in Britain."[49] When they arrived in New Zealand, war brides found outdated codes of dress that accentuated the gap in fashion between postwar Britain, where style had "moved on," and New Zealand, which was "stuck in the 1930s." This illustrates something of the clash between the metropolis and the periphery that was the remnants of the British Empire: war brides perceived that New Zealand did not match up to the Britain they had left behind, and curiously, New Zealanders perceived that war brides did not match up to the image that the conservative elements in that society perpetuated as "British" standards.

When the media championed war-bride marriages by extending desirable attributes to women including those from countries that had been wartime adversaries, it compounded the ambivalence of the public toward the brides. This was equally true when they overemphasized national stereotypes in describing the women. For example, in 1946, Mrs. Biddick, wife of an ex-POW, was deemed to be "unmistakably a German girl, her hair worn high in the German style with four combs, and youthful."[50] This report simultaneously drew attention to her German origins (possibly negatively received) and her youth and fashionable appearance combined with her willingness to learn English, which were positive features. This positive image was assisted by the fact that her family had sheltered the man she married when he escaped from a prisoner-of-war camp. British women came under the same scrutiny and were acclaimed for being demurely dressed, which implied acceptability.[51] In addition, they were praised for overcoming the strict rationing of clothes in Britain and arriving in New Zealand "all smartly turned-out."[52] Canadian and American brides and fiancées arriving in New Zealand on the *Monterey* in February 1946 were described in the press as "dressed quietly without any attempt to appear glamorous for the sake of effect."[53] However, images of neatly dressed constrained women were in competition with those of popular avant-garde fashions of war brides encouraged in the cinema newsreels and newspapers. War brides trod a thin line between presenting exciting new styles and being regarded as risqué.[54]

Just as the interest in fashion cast war brides as attractive "packages," the commodification of these women was compounded by the discourses of international "trade" in brides. There was a popular view that "war bride" was an epithet for prostitute and that they could be "acquired" by a simple exchange of goods.[55] The treatment of women who became war brides in this manner was encapsulated in a cartoon showing a New Zealand soldier about to be repatriated from Italy at the end of the war. The cartoon illustrated an Italian woman and a New Zealand private attempting to board the transport. The New Zealand soldier is addressing his commanding officer as he is about to be repatriated at the end of the war: "But Sir! Why can't I take her home? I hocked all my gear for HER instead of a CAMERA!" (figure 3.1).

In view of the fact that Italian women were portrayed in this way, it is not surprising that Italian war brides faced hostility when they arrived in New Zealand. However, senior members of the defense forces involved in the repatriation of war brides expressed the opinion that these women were making gigantic decisions to migrate to New Zealand, often in opposition to their families, and that they should be supported as much as possible.[56] This counterview was supported by letters to the editor of the *New Zealand Herald*. A returned serviceman wrote that "our admiration for the greater number of European girls was compelled by their bearing under great

Figure 3.1   This cartoon depicts a New Zealand soldier about to be repatriated at the end of the war addressing his commanding officer.

Source: Cartoon by Les Steel, printed in *The Tattler* (troopship magazine of the SS *Tamaroa*) (February/ March 1946), p. 11, Archives New Zealand/Te Rua Mahara o te Kāwanatanga, Wellington Office.

adversity, their loyalty to those separated by war from them, their dignity and many other virtues which, alas, did not seem to emanate in similar degree from so many back home here."[57] However, in spite of such positive protests on their behalf, the belief that women could be bartered for the price of a camera or the lure of a better life lingered. In the New Zealand context, Brigadier Stevens echoed these sentiments when he wrote in support of averting war marriages between 2NZEF personnel and Italian women:

> [I]t is probably true that the Italian people clearly recognise that after the war conditions of life will be infinitely more difficult in Italy than in New Zealand. Italian people have been very prone to emigrate at any time, consequently there is no widespread dislike to leaving Italy.[58]

New Zealand's immigration policy gave pause to a government obliged to facilitate the repatriation of dependents but anxious to control the entry of "non-British stock."[59] In general, the New Zealand authorities expected that cultural integration would be easier for women from "appropriate" and "complementary" backgrounds to their New Zealander husbands. The government also saw the benefits of acquiring a self-selecting population of women compatible with the projected immigration-promotion policy envisaged for the postwar period. Tensions existed because such a selection included women of diverse ethnic and religious populations, especially from North Africa and the Middle East.

However, although the tensions never entirely evaporated, the rights of servicemen to marry whom they pleased were protected by law, and in the postwar period the emphasis was on reuniting and rebuilding strong family units regardless of the background of the wives.[60] In this context, then, there seems to have been a common acceptance that "typical English milk-and-roses complexions and fine, fair, silky hair" would make for quick integration into New Zealand society.[61] These attributes could be extended to women from various backgrounds. The *Weekly News* reported that twenty-year-old Helene Bryant from Cairo had "the same kind of Greek beauty as the Duchess of Kent: oval face, brown eyes and soft, light brown hair."[62] Being "fair" was a visible signifier of the common British heritage the New Zealand government and community at large were keen to foster and indulge in as a part of their "national" identity. The woman who did not visibly disrupt this Anglo-Celtic imagined community was far more likely to be perceived as "bride" than whore.

Becoming a "Kiwi"

War brides and their children were "repatriated" in a crisscrossing pattern around the world. The term repatriation had in itself connotations that impacted on war brides' conception of themselves and of what they might expect from life in New Zealand. Being "repatriated" reinforced the notion that they were New Zealanders returning home, although they had never been to New Zealand before. On a personal level, war brides interpreted this to mean that they were not immigrants, marking out their privileged position on the basis of their marriage, and distinguishing them from other government assisted migrants arriving in the postwar period.[63]

Marriage transformed war brides into New Zealanders, and it was as New Zealanders that they undertook their voyages across the globe.[64] Upon marriage they automatically acquired New Zealand citizenship and the concomitant rights of entry and free passage to which dependents of servicemen were entitled.[65] Nancy Cott has argued that the conferring of citizenship privileges by way of marriage defines the boundaries of the nation.[66] However, on arrival in New Zealand, war brides discovered that they were not embraced wholeheartedly by the nation that had bequeathed them citizenship. A quest for acceptance and a sense of belonging then stretched out in front of them. They occupied contrary positions—supposedly being New Zealanders and at the same time not being familiar with local ways. They were required to relinquish their own customs and identities, but hand in hand with these expectations was a certain amount of acclaim for novelty. Being expected to fit in and simultaneously being regarded as different was sometimes difficult to accommodate, and many women endured ongoing rebuffs and struggled to attain equilibrium.[67] British-born Nancy Dickie wrote in 1988, 40 years after she arrived:

> My first struggle was trying to become a New Zealander. After all I was married to one and this country was to be my home so it was up to me to become a Kiwi. This challenge failed forlornly, trying to measure up to what I saw as the required standards was exhausting and unrewarding—time and common sense soon changed my mind. I decided that if I was to be happy I would be "English and proud of it" to the end of my days.[68]

Tsuruko Lynch, a Japanese war bride, expressed a similar sense of discomfiture. She was helped by neighbors to adapt to New Zealand society and "focus on being a good wife and mother."[69] Pressure to assimilate "meant that until quite recently [1999] she had put her life in Japan completely behind her."[70] It seems that this was not peculiar to Japanese women as Englishwomen found acceptance and acclimatization took time and were often not completely achievable. As Nancy Dickie found, rather than assimilate completely, she simply came to terms with her own "difference."[71]

In general, war brides were anxious to make a good impression on their in-laws and to avoid conflict and were, therefore, unlikely to object, as evidenced by English-speaking women who felt similar pressure to conform.

For example, religious affiliation proved to be a site of contention as families tried to impose their denominational beliefs on the newly arrived brides. Sylvia Smith, an Anglican from Derbyshire (United Kingdom), was expected to join the Methodist Church in Rangiora, the small South Island hometown of her in-laws. Her father-in-law presented her with a Methodist hymnbook soon after she arrived and, impervious to her protests, assured her that, having married his son, she was now a Methodist.[72] She experienced the letting go of her religious affiliations as a threat to her identity. In this case, English-speaking Sylvia felt that her past had to be abandoned if she was to be accepted in New Zealand.[73] Her compromise was to incorporate attendance at both churches each Sunday for as long as she lived with her in-laws.

War brides faced issues relating to acculturation and conforming regardless of their origins. Being labeled as "repatriated" and gaining citizenship did not automatically equate to being a "Kiwi." Differences in language, culture, and lifestyle brought war brides into sharp relief against their receiving communities. While they struggled to rationalize their choice of marrying a foreign serviceman and negotiated the transition to their husband's home country, they also had to overcome hurdles that the acquisition of citizenship privileges did not automatically level.

On migrating to New Zealand, gendered expectations dictated that they would be good wives and mothers, but they often found their in-laws skeptical of their capacity for the job—a skepticism based on suspicion of anything "foreign." British women, no less than European and other women, found this to be the case. Marygold Miller, English war bride, wrote to her mother in October 1946 that she was

> beginning to realize there was a lot of local resentment against the English which I did not expect. Here and there small slightly insulting remarks were passed which I did my best to ignore. I was amazed to find that English women were judged to be inefficient housekeepers, and every mistake I made was pointed out as proof of this.[74]

For the most part, war brides were willing to learn new skills and adapt to new living arrangements, ingredients, and cooking facilities that differed greatly from those they had previously enjoyed, but mistakes were lighted upon as evidence of ineptitude or "inefficient foreign ways."[75]

To assist war brides' adaptation, women were introduced to the New Zealand way of life, its culture, and customs through talks and films on board ship and in Italy even before they sailed. Instruction courses notwithstanding, war brides found themselves on a steep learning curve once they arrived in New Zealand. The government seemed anxious to promote conservative domestic and family values, distributing copies of Frederick Wood's *Understanding New Zealand* and *New Zealand: The Land, the People, the Way of Life*. In *Understanding New Zealand*, war brides could read about the "typical" New Zealand dairyman:

> The farmer who merely supervises the work of others is virtually unknown. He himself (often aided by his wife) splashes through the mud of Taranaki or the Manawatu or the Waikato to the cowshed.[76]

Professor Wood at least warned about the mud, and his books left them in no doubt of the rural nature of New Zealand society, the predominance of employment in primary production, and the sparseness of the population.[77] However, most war brides were unprepared for the demands of New Zealand life and could not always hide their disappointment at the lack of shops and entertainment.

War brides in New Zealand were relegated to the "home front," but in comparison with New Zealand women, most had a heightened awareness of what it meant to be at war. Home front and warfront had not, for many of these women, been separated by geography and for some had been one and the same. They found themselves out of step with the wives and mothers who had remained in New Zealand for the duration of the conflict and who had not had a similar "war experience." While New Zealand had actively participated in the war by supplying 130,000 Allied troops for war service and been a hub for manufacture and primary production, supplying Britain as well as the U.S. forces stationed in the Pacific, it did not suffer the immediacy of war as had occurred in Europe, the Middle East, and, arguably, North America.[78] Although New Zealand feared a Japanese invasion, this did not eventuate, resulting in a general lack of appreciation after the fact of the impact of war destruction experienced by many of the war brides. Talk of such experiences was deprecated, and war brides felt forced to drop their war (and other) past(s) from conversation. As noncombatants there was no place for "heroic" versions of their wartime experiences, but there was also little tolerance for what Penny Summerfield terms a "stoic" narrative version of their war lives.[79] While they may not have been combatants in the war that spawned their marriages, many had firsthand experience of bombings, blackouts, evacuation, severe shortages, and crossing oceans where torpedoes and mines were present. War brides arriving in New Zealand were taking up residency among a female population that had remained at the peripheries of the war and were faced with creating a truncated personal narrative—one that began the day they reached New Zealand.

Because they "belonged" in existing established families, war brides were quickly immersed in New Zealand society, possibly more quickly than other immigrants. Quick absorption is more problematic that it might appear. Anne Imamura noted that women who arrived as migrants in their husbands' countries could not move into "local society *gradually* but face immediate problems of establishing themselves in their husbands' social world."[80] It was at this point that war-bride clubs proved most useful to those women living in close proximity to one another. Based on their common experience of traveling on the ships and the mixed reception upon arrival, they offered one another acceptance, confidentiality, and moral support. Many war brides, however, were dispersed around the country to their husbands' rural homes, and the impact of this dispersal heightened their loneliness.[81] Although the war ended in 1945, the sense of dislocation experienced by these "war" brides extended far beyond the immediate end of the war and relocation to a peacetime society so very far away.

Useful Assets

Representations of war brides were an interesting mixture of commerce and domesticity. Side by side with gendered representations of war brides destined for domestic and child-rearing futures were welcoming speeches incorporating trade images made to war brides and delivered against a backdrop of Plunket nurses providing care facilities on the arrival wharfs for infants and their mothers.[82] Just as war had, for many, quite literally blown apart the walls of the domestic sphere and military and government officials intervened in personal relationships, the domesticity so longed for in the postwar era also exemplified the blurring of the private and the public. Housekeeping and child-raising abilities were scrutinized by receiving families backed up by the Red Cross and Plunket Society, a voluntary organization for the improvement of the health of mothers and infants.[83] Newly arrived war brides were introduced to the Plunket philosophy at the earliest opportunity, and photographs of Plunket nurses showing off the "new New Zealanders" were regular features in the press.[84] Inexperienced in infant care and bereft of family support, war brides made good use of Plunket nursing services although conflict sometimes resulted because of the disparity between routines promoted by the Plunket Society and war brides' own heritage.[85]

Some war brides found that culturally specific knowledge about child-rearing and wifely roles did not transplant well.[86] The rearing of children according to the strict "Plunket way" went against the ethos of those whose cultural norms favored a more inclusive nurturing style of infant care.[87] Or, as an English bride found, it was possible to get on the off-side with the

Plunket nurse merely by purchasing a feeding formula other than that recommended by the Society.[88] War brides appear to have coped with the mandates of the Plunket nurse in similar ways to local women—when they disagreed with Plunket they simply ignored the advice given or ceased attending Plunket clinics. However, whereas local women could often fall back on family or friends for advice, war brides faced pregnancy and child-rearing without natal family support and were more isolated in this respect. The lack of infrastructure and support networks of their own kin was sorely felt. In her story of caring for her firstborn child, John, Kath Adams encapsulated how the absence of natal family affected her: "I can remember John howling and howling. He's just six weeks younger than Prince Charles. And I said, at least the Queen doesn't have to put up with this! Someone [in the Royal household] would know what to do, and I didn't."[89]

Although she could not, or would not, recall homesickness per se, there is poignancy in Kath's account of having to cope on her own. This was a common reaction among war brides. Their sense of alienation and dislocation was often most vividly described in connection with pregnancy, infant care (and death), and housekeeping, especially cooking and keeping the cantankerous wood-burning ranges alight. The gendered nature of their roles as housewife and mother served to suggest that they could perform these tasks irrespective of location, whereas the reality was that the absence of familial support networks deprived them of vital, culturally specific, and familiar routines.[90] New Zealand's war brides often faced extreme hardships, especially initially, because of housing shortages, old-fashioned cooking facilities, and lack of household appliances. They knew that their lack of skills in coping with New Zealand conditions was a matter of conjecture and recognized that they were viewed as falling short of the ideal. It was at this point that mothers-in-law, intolerant of their failings, often openly used derogatory terms linking the newly arrived bride in no uncertain way with brothels and loose living. There was a concerted effort to present incoming war brides, including those from former enemy territories, as attractive, feminine, mastering the English language, and as capable of discharging their duties as mothers. But running parallel with efforts to portray war brides as ideal wives and mothers was a pervasive idea that they were morally suspect, sexually available, and inept housekeepers.

Conclusion

New Zealanders portrayed the women who became the war brides of "their" men in contradictory terms. Publicly, they were acclaimed as ambassadors, model settlers, and contributors to the "balancing of the books" in terms of imports and exports. Besides being assessed in commercial terms, they were judged on moral, demographic, and logistical criteria, and their perceived attributes and deficiencies were a matter of public debate. The media scrutinized their origins, appearance, and demeanor, portraying them as harbingers of modern fashions, adaptable, and compatible with New Zealand society while simultaneously drawing attention to their risqué behavior and possibly dubious motives for marrying New Zealand servicemen. Privately, in-laws examined and questioned war brides' capacity to be good wives and mothers and were variously intrigued and scandalized by their clothes and habits.

The war experiences and service of the war brides marked them off as dissimilar to the majority of local women and were completely overshadowed by the war service of New Zealand's returning veterans. On arrival in New Zealand, war brides faced prejudice and conservative attitudes as well as well-meaning, if sometimes narrow-minded, kindness and acceptance. War brides were, even in their own eyes, at once privileged and disadvantaged. They were privileged by virtue of their marriage to a New Zealand serviceman and the rights to free passage and entry to the country that it implied; they were disadvantaged by being "foreign" and having to face the journey and life in New Zealand without family support. Administrative procedures and gendered expectations categorized them in ways that predetermined their roles and directed them to fit into the receiving society intent on regenerating the family and reigniting the national economy. War brides' family histories, their war service, and life experience were subsumed by marriage, motherhood, and domestic life. In the process of becoming a war bride and traveling to New Zealand, these women found their dependent status firmly established and their occupation in the domestic arena assumed.[91]

New Zealand's sample of incoming war brides demonstrates how moving between countries as a result of wartime encounters impacted on these women. Government policies relative to citizenship, nation building, and maintenance of the British Empire, economic, and commercial considerations, all contributed to the reconfiguration of war brides' identity and sense of belonging in the postwar period. The women themselves have overwhelmingly described the impact of arrival as heightened by the sense that they needed to abandon their past affiliations, customs, and histories. The full glare of publicity that greeted war brides on arrival focused attention on their compatibility with New Zealand culture and society or, at least, their willingness to adapt to it. Signs of difference were muted as press reports sought to convey the impression of easily assimilated women and children. At the same time, lingering prejudices surfaced in families and communities critical of the perceived shortcomings especially in the domestic arena. Taken together, these issues contributed to conflicting messages being received by war brides, and, on the whole, women found New Zealand society contrived to obliterate their pasts (especially their war lives) and required them to conform to the status quo.

This is the story not of a few extraordinary women but of a mass movement of women from the mainstream of their societies. New Zealand's sample of war brides is colorful in its diversity, yet common themes run through their experience. Most found it difficult. As one war bride in New Plymouth said, "I can tell you categorically I wouldn't do it again."[92] This is not because she regrets her life in New Zealand. It is because she still remembers the acute pain of adjustment and difficulty in bridging the gap between her war (and peace) pasts and her chosen future in New Zealand. Whether welcomed as bride or regarded suspiciously as having disembarked from a brothel, war was the formative experience of these women. Women, who in the wake of—often quite literally—losing their former homes, struggled against the simultaneous loss of individual identities and the imposition of impossible ideals and damaging stereotypes.

Notes

1. War bride: "A woman who marries a soldier (especially a foreign one) during a war," in *Shorter Oxford English Dictionary*. 5th ed., vol. 2. New York, 2002, p. 3577.

2. Penny Summerfield, *Reconstructing Women's Wartime Lives: Discourse and Subjectivity in Oral Histories of the Second World War*. Manchester, NY, 1998, p. 13.

3. The unusual circumstances of war brides' marriages and migration have drawn a welter of attention from academics as well as novelists, memoir writers, and popular history writers. For example, Gerald J. Schnepp and Agnes Masako Yui, "Cultural and Marital Adjustment of Japanese War Brides," *American Journal of Sociology*. 61, 1, July 1955, pp. 48–50; Anselm Strauss, "Strain and Harmony in American-Japanese War Bride Marriages," *Marriage and Family Living*. 16, 2, May 1954, pp. 99–106; Petra Goedde, "From Villains to Victims: Fraternization and the Feminization of Germany, 1945–1947," *Diplomatic History*. 23, 1, Winter 1999, p. 17; Raingard Esser, "Language No Obstacle: War Brides in the German Press, 1945–49," *Women's History Review*. 12, 4, 2003, pp. 577–603; Annette Potts and Lucinda Strauss, *For the Love of a Soldier: Australian War Brides and Their GIs*. Sydney, 1987; Pamela Winfield, *Sentimental Journey: The Story of the GI Brides*. London, 1984; Teresa K. Williams, "Marriage between Japanese Women and US Servicemen since World War 2," *Amerasia Journal*. 17, 1, 1991, pp. 135–154; Paul Spickard, *Mix Blood: Intermarriage and Ethnic Identity in Twentieth-Century America* Madison, WI,

1989, p. 130; Suenaga Shizuko, "Good-Bye to *Sayonara*: The Reverse Assimilation of Japanese War Brides" PhD thesis, Boston College, Boston, 1996, p. 22; George de Vos, *Socialization for Achievement: Essays on the Cultural Psychology of the Japanese*. Berkeley, CA, 1973, pp. 248, 252; Keiko Tamura, "Border Crossings: Japanese War Brides and their Selfhood" PhD thesis, Australia National University, Canberra, 1999; Evelyn Nakano Glenn, *Issei, Nisei, Warbride: Three Generations of Japanese American Women in Domestic Service*. Stanford, CA, 1986; Bok-Lim C. Kim, "Asian Wives of US Servicemen: Women in Shadows," *Amerasia*. 4, 1, 1977, pp. 91–115; Elfrieda B. Shukert and Barbara Smith Scibetta, *The War Brides of World War 2*. Novato, CA, 1988; John Hammond Moore, *Over-Sexed, Over-Paid and Over Here: Americans in Australia 1941–1945*. St. Lucia, 1981; Jenel Virden, *Goodbye Piccadilly: British War Brides in America*. Urbana, IL, 1996; Harry Bioletti, *The Yanks Are Coming: The American Invasion of New Zealand 1942–1944*. Auckland, 1995; Jock Phillips (with Ellen Ellis), *Brief Encounter: American Forces and the New Zealand People 1942–1945*. Wellington, 1992; Carol Fallows, *Love and War: Stories of War Brides from the Great War to Vietnam*. Australia and New Zealand, 2002; Orapan Footrakoon, "Lived Experiences of Thai War Brides in Mixed Thai-American Families in the United States" PhD thesis, University of Minnesota, Minnesota, 1999; Lois Battle, *War Brides*. New York, 1982; Velina Hasu Houston, ed., *The Politics of Life: Four Plays by Asian American Women*, Philadelphia, 1993, pp. 205–274; Roberta Uno, *Unbroken Thread*. Amherst, MA, 1993, pp. 155–200; Helene Lee, *Bittersweet Decision: The War Brides Forty Years Later*. Lockport, IL, 1985; Barbara B. Barrett, Eileen Dicks, Isobel Brown, Hilda Chaulk Murray, and Helen Fogwill Porter, eds., *We Came From Over the Sea, British War Brides in Newfoundland*. Portugal Cove, NL, 1996; Vera A. Cracknell Long, *From Britain with Love, World War II Pilgrim Brides Sail to America*. New Market, VA, 1988; Monette Goetinck, *Bottled Dreams*. Napa, CA, 1998; Dorothy McCormack Graw, *A Heart Divided: A War Bride at Home in Two Worlds*. Bloomington, IN, 2004; Marygold Rix-Miller, *Trophy of War*. Bognor Regis, UK, 1983; Jeanne Molloy, *While I Remember: Memoirs of Jeanne Olwyn Molloy*. Auckland, 2001; Leicester Smith and Sylvia Smith, *Pathfinder: Wings in the Crowded Sky*. Auckland, 2004, MS2004/37, Auckland War Memorial Museum Library, Auckland, New Zealand (AWMM); George J. Sanchez, "Race, Nation, and Culture in Recent Immigration Studies," *Journal of American Ethnic History* 18, 4, Summer 1999, pp. 66–86; Caroline Chung Simpson, "'Out of an Obscure Place': Japanese War Brides and Cultural Pluralism in the 1950s," *Differences: A Journal of Feminist Cultural Studies*. 10, 3, 1998, pp. 47–81; Regelio Saenz, Sean-Shong Hwang, and Benigno E. Aguirre, "In Search of Asian War Brides," *Demography*. 31, 3, August 1994, pp. 549–559; Debbie Storrs, "Like a Bamboo: Representations of a Japanese War Bride," *Frontiers*. 21, 1/2, January 2000, pp. 194–224.

4. John Costello, *Virtue Under Fire: How World War 2 Changed Our Social and Sexual Attitudes*. Boston and Toronto, 1985.

5. *New Zealand Herald.* (NZH) 25 July 1945, p. 9.
6. Army Secretary, Memorandum, 6 December 1945, D334/4/1/A/2, Archives New Zealand/Te Rua Mahara o te Kāwanatanga, Wellington Office (ANZW).
7. N. H. Carrier and J. R. Jeffery, *Studies on Medical and Population Subjects, No. 6, External Migration: A Study of the Available Statistics 1815–1950.* London, 1953, pp. 38–41, 54.
8. Gabrielle A. Fortune, "'Mr Jones' Wives': World War II War Brides of New Zealand Servicemen" PhD thesis, University of Auckland, 2005.
9. HQ 2NZEF Cairo to HQ Wellington, Telegram, 29 January 1944, AD1 320/4/4 Pt. 3, ANZW.
10. *Weekly Review.* (WR) no. 190, New Zealand Film Archive, Wellington (NZFA).
11. *Taranaki Herald.* (TH) 16 February 1946; *NZ Truth.* 2 July 1947, p. 9.
12. NZH, 20 March 1946, p. 8.
13. Ibid.
14. Press cutting, 18 July 1946, Val Wood Archive (VWA), Hamilton Public Library, New Zealand (VWA).
15. Robin Kay, *Official History of New Zealand in the Second World War 1939–45: Italy, From Cassino to Trieste,* vol. 2. Wellington, 1967; Ingrid Bauer, "'Austria's Prestige Dragged into the Dirt'? The 'GI-Brides' and Post War Austrian Society (1945–1955)," *Contemporary Austrian Studies.* 6, 1998, p. 43.
16. "If the Cap Fits" WAII 1DA 426/29–40, ANZW.
17. Ibid.
18. D. H. Davis, *Soldier's Guide to Italy.* p. 13, MS2002/75, AWMM.
19. Bauer, "'Austria's Prestige Dragged into the Dirt,'" p. 43.
20. Bok-Lim C. Kim, "Asian Wives of US Servicemen: Women in Shadows," *Amerasia Journal.* 4, 1, 1977; Virden, *Goodbye Piccadilly;* Nora Smith, interview, February 2002.
21. Susan R. Grayzel, *Women's Identities at War: Gender, Motherhood, and Politics in Britain and France during the First World War.* Chapel Hill, NC, 1999, p. 122.
22. SS1/616, ANZW.
23. TH 4 January 1946, 10 January 1946, 11 January 1946.
24. Melbourne *Age.* 5 January 1946.
25. Ibid., 23 March 1946.
26. Ibid., 10 August 1946.
27. NZH 14 January 1946, p. 4.
28. E. J. Williams to Viscount Addison, 27 June 1946, DO35/1/98 Z484/1/2, National Archives, Kew, London.
29. Sylvia Smith, interview, 8 August 2001.
30. Generally, husbands and boyfriends were repatriated separately and in advance of their wives and fiancées.
31. NZH 12 January 1946, p. 6.
32. "Horse races" on troopships consisted of horses (war brides in this instance) carrying cardboard cutouts and moving along the "course" in response to the score on a dice thrown by the jockeys. The punters placed bets on their "horse."

33. Press cutting, 18 July 1946, VWA.

34. A. Trevelyan, "All at Sea," *Stirling Castle* Newsletter, 1946. WAII 1DA 453/50, ANZW.

35. BBAE A304/831/149, ANZA.

36. Fortune, "Mr Jones' Wives."

37. Deborah Montgomerie, *The Women's War: New Zealand Women, 1939–45.* Auckland, 2001, p. 187.

38. Margaret R. Higonnet, Jane Jenson, Zonya Michel, and Margaret C. Weitz, "Introduction," in Margaret R. Higonnet, Jane Jenson, Sonya Michel, and Margaret C. Weitz, eds, *Behind the Lines: Gender and the Two World Wars.* New Haven and London, 1987, p. 4.

39. F. L. W. Wood, *Understanding New Zealand.* New York, 1944, p. 97.

40. Joan Wallach Scott, *Gender and the Politics of History.* New York, 1988, p. 42.

41. Martha Gardner, *The Qualities of a Citizen: Women, Immigration, and Citizenship, 1870–1965.* Princeton, NJ, 2005, p. 225.

42. Eileen Haughey, interview, 22 August 2001; *Weekly News.* (WN) 10 January 1945, pp. 12–13; WN 26 April 1944, p. 9; WN 31 May 1944, p. 10; *Press.* 14 January 1946, p. 4.

43. Lucy Noakes, *Women in the British Army: War and the Gentle Sex, 1907–1948.* London and New York, 2006, p. 16.

44. For a discussion of elevated position of combatants: Noakes, *Women in the British Army,* pp. 1–19.

45. Single women (fiancées) were more likely to have their occupation recorded on the ships manifesto. Married women (wives) were more likely to be listed as occupied in "Home Duties."

46. WN 10 January 1945, pp. 12–13; WN 26 April 1944, p. 9; WN 31 May 1944, p. 10; *Press.* 14 January 1946, p. 4.

47. BBAO5552/35, ANZA.

48. *New Zealand Women's Weekly.* 15 March 1971, p. 34; press cutting, VWA.

49. Jeanne Molloy, interview, 19 April 2001.

50. *Press.* 14 January 1946, p. 4.

51. Noakes, *Women in the British Army,* p. 74.

52. WN 26 April 1944, p. 9.

53. NZH 9 February 1946, p. 6.

54. NZH 14 November 1946, p. 4.

55. Noakes, *Women in the British Army,* pp. 1–2.

56. Telegram from Minister of External Affairs, Wellington, to Acting High Commissioner for New Zealand, London, 2 April 1946, EA1 95/6/5 Pt. 2, ANZW.

57. Press cutting, M. P. McDermott private collection.

58. Brigadier W. G. Stevens, OIC 2NZEF Italy to Army HQ, Wellington, 31 December 1943, AD1 320/4/4, ANZW; Memorandum, Army HQ to Minister of Defence, 3 July 1944, AD1 320/4/4 Pt. 2, ANZW.

59. P. S. O'Connor, "Keeping New Zealand White, 1908–1920," *New Zealand Journal of History.* (NZJH) 2, 1, 1968, pp. 41–65; Sean Brawley, "No 'White

Policy' in NZ," NZJH 27, no. 1 (1993): 16–36; Immigration Restriction Act, 1908 and amendments; Kamy Ooi, "The Liberalization and End of the White New Zealand Immigration Policy, 1946–1987" MA thesis, University of Auckland, Auckland, 1999.

60. A. E. Currie, Crown Solicitor to Adjutant-General Army HQ, 15 March 1944, refers to opinions issued by the Crown Law Office 22 October 1940; 12 May 1943; 15 March 1944, WAII 1DA 6/9/30, ANZW.

61. WN 26 April 1944, p. 9.

62. WN 10 January 1945, p. 12.

63. Fortune, "Mr Jones' Wives," p. 194.

64. Malcolm McKinnon, *Immigrants and Citizens: New Zealanders and Asian Immigration in Historical Context*. Wellington, 1996, p. 36.

65. In this respect war brides to New Zealand were better served than their counterparts to other countries. They were guaranteed entry and travel costs. For a discussion of the rights of war brides entering the United States, see Gardner, *Qualities of a Citizen,* especially pp. 224–227.

66. N. Cott, *Public Vows: A History of Marriage and the Nation*. Cambridge, 2000; Beatrice McKenzie, "Gender and United States Citizenship in Nation and Empire," *History Compass*. 4, 2006, pp. 592–602.

67. Rix-Miller, *Trophy of War*. Bognor Regis, UK, 1983.

68. Nancy Dickie, August 1988, MS Box 0113, VWA.

69. Tsuruko Lynch, interviewed by Peter Boston, 21 October 1997, Auckland, as quoted in Peter Boston, "Tsuruko Lynch: From Shike to Northcote," in Roger Peren, ed., *Japan and New Zealand: 150 Years*. Wellington, 1999, p. 148.

70. Peren, *Japan and New Zealand*.

71. Dickie, VWA.

72. The Smith family disapproved of married women in paid employment, so Sylvia's desire to get a job was also thwarted.

73. Sylvia Smith, interview.

74. Rix-Miller, *Trophy of War*, p. 173.

75. Audrey B. King, "The Foreigner," YWCA Review. October 1947, pp. 14–15.

76. Wood, *Understanding New Zealand*, p. 97.

77. High Commissioner for New Zealand, *New Zealand: The Land, the People, the Way of Life*. London, c. 1945. Secretary of External Affairs to New Zealand High Commissioner, Ottawa, 16 July 1945, EA1 63/4/1, ANZW.

78. For details of New Zealand war casualties, and food exports, see http://www. NZHistory, http://www.nzetc.org/tm/scholarly/WH2Econ-fig-WH2Eco066a. html, and http://www.nzetc.org/tm/scholarly/WH2Econ-fig-chart65.html. Accessed December 2007.

79. Summerfield, *Reconstructing Women's Wartime Lives*.

80. Anne E. Imamura, "The Loss That Has No Name: Social Womanhood of Foreign Wives," *Gender and Society*. 2, 3, September 1988, p. 292. My emphasis.

81. Maria A. Rivera, Mary Nash, and Andrew Trlin, "Here I am Everyone's Umbrella: Relationships, Domesticity and Responsibilities—The Experiences of

Four Latinas in New Zealand," *Women's Studies Journal.* 16, 1, Autumn 2000, p. 56; Teresa K. Williams, pp. 135–154.

82. WN 26 April 1944, p. 9.

83. Margaret Tennant, "History and Social Policy: Perspectives from the Past," in Bronwyn Dalley, Margaret Tennant, eds., *Past Judgement: Social Policy in New Zealand History.* Dunedin, New Zealand, 2004.

84. WN 26 April 1944; WR No. 190, NZFA.

85. Sally Giffney, interview, 11 September 2001; Mirella Hall, interview, 19 June 2001; Thelma Roberton, interview, 25 July 2001.

86. Imamura, "Loss That Has No Name," pp. 291–307.

87. Peter Blunden, interview, 5 December 2001.

88. Mary Emmett, interview, 1 October 2000.

89. Kath Adams, interview, 22 July 2001.

90. Regarding the difficulty of transporting "womanhood" from one international location to another, see Imamura, "Loss That Has No Name," pp. 291–307.

91. F. L. W. Wood, *This New Zealand* . . . Hamilton, UK, 1946.

92. VWA.

CHAPTER 4

Innocence and Punishment: The War Experiences of the Children of Dutch Nazi Collaborators

Ismee Tames

The German invasion of the Netherlands on 10 May 1940 made the anonymous author of *Onder de vleugels van de partij: kind van de Führer (Under the Wings of the Party: Child of the Führer)* feel excited.[1] The invading Germans were National Socialists just like himself and his parents. Now, finally, the bullying and street fights between the few supporters and many adversaries of National Socialism would be over. The new era had arrived! But living in Rotterdam, the 12-year-old soon saw his city erupt in flames during *Luftwaffe* bombing four days later. The adventure that was supposed to come with war was accompanied by fear and anxiety and fundamentally changed his life. In the course of the war, he would be sent to a National Socialists-children's home in eastern Germany, conscripted into the *Wehrmacht,* and eventually enter the service of the SS. This Dutch boy ended the war actually fighting on the German side in the battle of Arnhem and in the Ardennes.

In contrast to this story, most children of Dutch National Socialists remained noncombatants and did not experience the harsh realities of war and occupation until 5 September 1944, when the Allies liberated the southern provinces of the Netherlands and collaborators fled en masse out of the recently liberated areas. Dolle Dinsdag (Mad Tuesday), as 5 September came to be known, plays an important role in the autobiographies and memoirs that these children later wrote.[2] Throughout the course of September and October, collaborators who had not succeeded in getting north of the main rivers were mostly arrested. The rest of the Netherlands remained in German hands and experienced the harshest period of the Nazi occupation. For many collaborators, the period from September 1944 up to the summer of 1945 was a time of growing insecurity and degradation. This period was particularly harrowing for the women and children involved, who, in spite of often having been minimally politically active, were now faced with the far-reaching consequences of the political choices and deeds of their husbands and fathers.

"Suffering" was from the beginning regarded as an integral element in the Dutch experience of the Second World War, epitomized by the idea that this small neutral country had been brutally invaded by its large, aggressive neighbor and had been subjected to its regime of terror for five long years. Suffering and innocence are combined themes in Dutch postwar national identity,[3] having become synonymous with rejecting evil and, therefore, with being on "the good side."[4] By the 1960s, the idea of national suffering was superceded by the suffering of various groups within society: the Dutch Jews were the subjects of initial attention but were soon followed by other groups affected by the war, such as bombardment victims, resistance fighters, and the "second generation," including the children of collaborators. The definition of suffering at the hands of the German enemy or National Socialism broadened to include suffering endured as a result of decades of societal ignorance and neglect or, rather, the suffering inflicted by postwar society on "war victims." Children in testimony books thus became, in their own minds and through their publications, representatives of the helpless innocent nation that could not withstand the horrors of war.[5]

The memoirs or "testimony books" of children of collaborators began to appear in the early 1980s and drew attention to the effects of the war and occupation on them at the time and later in their adult lives.[6] The children, both those who actively contributed to the German war effort and those who did not, present their stories in these memoirs within the narrative framework of the "innocent child." In doing so, they mobilize the notion of unjust suffering that is central to the Dutch collective memory of the war, from which they are otherwise largely excluded because of their status as the children of the "enemy within." Placing themselves under the same umbrella as other "war victims" had obvious therapeutic effects on the people concerned. However, the significant peculiarities of their actual individual experiences and memories must be noted in examining their testimonies.

Unlike other "war children," the offspring of collaborators shared a diffuse fear of a Bijltjesdag (a Day of Reckoning): an Allied victory was expected to unleash the revenge of the Dutch population on the collaborators. Collaborators' children often hoped for a German victory or at least a "draw."

These hopes, fears, and expectations of a postwar future clearly set these children apart from the rest of the Dutch population. In their memoirs, children of collaborators blur the distinctions between the concepts of "victim," "perpetrator," and "bystander" and thereby invite a questioning of the divisions between "right" and "wrong" (*goed* and *fout*) that were forcibly imposed by the Dutch on all who lived in the country during the war. During the 1980s, however, the accounts of these children began to attract increasing public interest following a critical shift in attitudes away from the early postwar myths of national heroism and resistance in the Netherlands.[7] This interest was sharpened in conjunction with a move in scholarly and public interest from heroism to victimhood. The image of the innocent child who had endured hardships replaced questions regarding the responsibility of perpetrators of war crimes and issues of guilt and responsibility that had previously dominated Dutch views of those who had in one way or another collaborated with the Nazis.

Telling the war stories of collaborators' descendents who return to the position and perspectives of childhood in their memoirs has important consequences. In Western culture, especially since the twentieth century, childhood is regarded as a distinct phase in life. Childhood is also regarded in a utopian light: it ought to be carefree and a time of happiness. When a child "loses" his or her innocent childhood, it is generally perceived as unfair or even tragic.[8] These cultural notions about how childhood should be experienced influence how readers interpret stories about children. When the main protagonist of a war story is a child, the narrator has the possibility of addressing questions that would be much more difficult if told from an adult's perspective. For instance, the notion of helplessness associated with children allows the helplessness of adults to be addressed—a much more uncomfortable and problematic reality but one of crucial importance when thinking about the impact of war on combatants and noncombatants alike.

It is also important to interrogate the cultural association of helplessness with innocence. The supposed helplessness of the child invites us to regard him or her as an innocent who cannot, therefore, be interrogated on political or moral issues. "Innocence" makes every evil that befalls the child seem automatically "unjust" or "unfair."[9] When collaborators' children present their war stories first and foremost *as children,* the first impulse of the reader to focus on the deeds and ideas of the parents or the moral choices of youngsters engaged directly with the war effort is diverted. The adoption of a "childlike persona" makes it possible for the author to avoid the larger political and historical context of their stories and remove the burden of being scrutinized as a Nazi supporter.

Many of the memoirs written by the offspring of Dutch collaborators focus on exclusion and societal revenge, both feared and real, and on all who were stained by National Socialism. The framework of the "innocent child" thus begins by underscoring an injustice perpetrated by society: these children are depicted as the innocent victims not of their parents' choices but of society's blind revenge. In some memoirs, parental choices are questioned; in others, they are not. This does not mean, however, that for all children of collaborators this framework of punishment by society functions in the same way. As this chapter demonstrates, retribution and the way it is remembered depend on the specific experiences of these children in the last stages of the war. The kinds of "social punishment" they identify depend upon whether they remained at home, were evacuated to rural provinces, or ended up in Germany. The memoirs of children of collaborators enhance our understanding of the ways in which childhood memories are framed. They can reveal the means of dealing with the legacies of collaboration and how these memories have become cohesively integrated in the Dutch collective memory of the Second World War.

The Children Who Stayed

During the occupation, NSB members constituted a small minority of the Dutch population. They were generally seen as traitors to the national cause, supporters of an "un-Dutch" ideology and of the Nazi dictatorship and its regime of terror.[10] The Germans were, of course, the real enemy, but followers of the NSB were regarded as the "enemy within," a position even more worthy of contempt and disgust. The children of NSB members were equally abhorrent in the eyes of their contemporaries, all the more so because Nazi propaganda presented them as the "new generation," who would build the National Socialist future in the Netherlands. To this end, they were frequently dressed in the uniform of the Nationale Jeugdstorm (National Youth Storm, NJS) at official gatherings of the NSB, at meetings with the occupying forces, or at visits of high-ranking Nazis from Berlin. The wartime NSB-controlled press commented on sports exchanges between the NJS and Hitler Jugend (HJ), while young males were called up to fulfill their duty against the Soviet Union. Regularly, groups of children could be seen assembling at railway stations to go to holiday camps in Germany, Austria, or the Sudetenland. Even closer to home, they sold NSB newspapers, raised money for National Socialist welfare organizations, and marched through the streets singing about the imminent National-Socialist era. To the Dutch public at the time, these children were a constant reminder of the Netherlands' weakness and national humiliation in the face of the Nazi occupation. The children were symbols of the ongoing occupation and one of the most troublesome representations of the "enemy within." Predictably, they were often isolated and bullied by the rest of the Dutch population.

Unsurprisingly, among collaborators and their families, there was a constant fear of a Bijltjesdag. Neighbors, colleagues, or schoolmates whispered that they would take their revenge when the liberation came and that Dutch Nazis would be "swinging from the highest trees."[11] This of course frightened the children, even though many of their parents insisted that a German victory was imminent and that they should not pay any attention to their frustrated schoolmates. In spite of parental reassurances, some children still had nightmares about Bijltjesdag. Hendrik, who was eight when the war broke out, expected that the world would be turned upside down if the Germans lost the war. He feared that the signs that now read *Juden nicht erwünscht* (Jews not wanted) would change into *NSB-ers niet gewenscht* (NSB-ers not wanted) and that they would all be forced to wear swastikas on their clothes.[12] While this never happened, it does demonstrate the existence of fears about a day of vengeance long before liberation came.[13]

Sytze van der Zee's memoirs exhibit similar preoccupations when he writes that his parents left the NSB long before the war's end hoping to salvage their reputations.[14] The damage had already been done, though. The fact that his father sold his WA (Weerafdeling, NSB militia group) combat uniform after he ended his membership did not make his neighbors forget that he had once possessed it. When Dolle Dinsdag came, Sytze and his siblings were sent to their grandparents in the countryside. When nothing happened, the children soon returned. Their parents had, in the meantime, begun a wild hunt for any incriminating documents that then had to be destroyed.[15] This chaos within the domestic sphere mirrored what was happening on the streets. Van der Zee describes his experiences as a six-year-old in an increasingly devastated bourgeois neighborhood during the last winter of the war. The young Van der Zee tried to assert his place as a part of the neighborhood where everything revolved around finding wood, coal, and food.

Van der Zee's memoirs, *Potgieterlaan 7*, were published in 1997 and received considerable media attention.[16] During the 1980s, Van der Zee was chief editor of *Het Parool*, a newspaper that had been founded illegally during the occupation and still had a strong identification as a "resistance" publication. People were, therefore, surprised to find out that he was descended from an NSB family. Van der Zee's secret only became public knowledge when he published his memoirs after the deaths of his parents and eldest brother. He waited until then to shield them from negative reactions to their past. In this way, the "innocent child" became his parents' protector. In his book, Van der Zee moved between past and present, examining his childhood and its effects on his later life and particularly his relationship with his parents. During and shortly after the war, he was known in the neighborhood and at school as the son of an NSB member. The other children called him and his brother "filthy NSB-ers" and "traitors," but at the same time, Van der Zee describes how they shared the other children's fear of a real fanatic and authoritarian NSB man who lived close by.[17] Although Van der Zee was teased by his peers, he was also a member of their street gang and took part in fights with children from other neighborhoods. When liberation finally came, the Van der Zee children were nevertheless excluded from the liberation parties in the neighborhood. Amid rumors about people being arrested, the family sat at home and waited for something to happen. Finally, soldiers came to the door and arrested his father at gunpoint. This, naturally, made a huge impression on the rest of the family and especially the mother, who was also desperately afraid of being arrested. The family went underground in their own house, hoping that people would assume they had left. The house and the household deteriorated. After a while, news from his father came and they went to visit him in a nearby camp. Sytze thought his father looked quite normal and made a calm impression, although his head had been shaved. The camp was not too bad, his father said. He explained the theatrical arrest by the armed soldiers as a case of mistaken identity. The soldiers believed he was another man who was also called Van der Zee, who had betrayed Dutch Jews to the Germans.[18]

Sytze's mother seemed far more disturbed by what had happened than his father. After his arrest, she was on her own in a hostile environment and with an entire family to feed. She began to lose control. She cried, was suicidal, and fought constantly with Sytze's eldest brother. Sytze hung around on the streets with the other children where he more or less belonged to the group. Still, for this young boy, the well-known, trusted neighborhood could suddenly become a hostile environment where his father could be taken away and he could be bullied. The gap between his experience and that of the other children in the street is clearly described in his book when he observes how he saw the city's recovery from the war: the garbage was collected again, the streets were rebuilt, the houses painted. It was only his house that gradually began to crumble away. Like the house, his family had fallen to pieces and seemed to recover far more slowly than everyone else's.

The dual face of their own community was felt even more directly by other children of collaborators. Sometimes NSB families were dramatically torn apart, and the arrests were accompanied by violence and the humiliation of the parents. Janny and Pia, for instance, recall beatings by locals and public humiliation,[19] as do Eva and Marrie.[20] Marrie was fifteen when her village was liberated. The local population besieged and pillaged her family house. Before being arrested and taken away by the local authorities, her parents were dragged into the street, where Marrie's mother's head was shaven and her father was beaten. In the end, Marrie was left alone with her little brothers and sisters until friends of her parents came to collect them.[21] Marrie's story stresses the punishment of these children who had not been "active" as collaborators, let alone as combatants, and were, therefore, not guilty of any wrongdoing. She framed their experiences as the cruel punishment of the innocent, not as a sad consequence of the deeds and choices of her parents.

The moment that their fathers, or sometimes both parents, were arrested is crucial in the recollections of many collaborators' children who experienced the liberation on the thresholds of their own homes. It was at this point that their trust in what they had assumed to be a known and safe world was eroded. Their stories often revolve around being threatened (men with guns or people yelling that they will shoot them) or actual physical violence (a father pushed down the stairs or beaten in the street).[22] The moment of arrest marked a point of tremendous insecurity for the children.

When studying instances of violence and punishment in the memoirs, it is important to point out how some of the testimonies were collected. Janny and Pia, for example, were interviewed about their experiences as a part of a student's research project in the late 1980s. The resultant thesis intended to "give a voice" to former children of collaborators who had negative experiences in the immediate postwar period. Eva and Marrie were interviewed over a decade later with a similar purpose, although this time the interview project was initiated by a popular historical magazine and resulted in a book.[23] The aim of both works was to promote stories and interests of the "silenced victims" of postwar reconstruction. The interviewees were invited to talk about their negative experiences but not necessarily examine what may have shaped their memories. This approach had a huge impact on what was recalled and presented in the subsequent publications. Negative experiences were regarded as the rule and the positive ones as exceptional. Many stories stressed moral outrage at a society that visited the crimes of the collaborator parents upon the "innocent" shoulders of the children.

The purpose of giving "a voice" to "a silenced group" may be therapeutically adequate under certain circumstances and may also facilitate public debate about the long-term consequences of war, but it is also a way of according primacy to some memories while marginalizing the impact of others. Narratives in which a violent arrest or humiliation of the parents did not occur or of children who were the recipients of relatives' or neighbors' hospitality came to be categorized as atypical and pushed to the margins. While both these collections reveal instances of "normal" or friendly behavior by others, they are presented as unusual in the quest to accord victim status to the children and to question national wartime mythologies.[24]

Different liberation experiences are connected with the fact that there was no consistent or clear policy on the arrest of collaborators. In some cities, the organized resistance had lists of NSB members who were required to be turned in. In others, personal scores were settled or it was disputed whether the resistance, Binnenlandse Strijdkrachten (Domestic Forces, BS), or the police were to do the arresting.[25] Sometimes the arrest developed into an "event" with crowds watching and shouting. Compared with France, for instance, only a few collaborators were actually killed. But when it came to the threat of violence, Bijltjesdag expectations seemed to be a confirmation of what many NSB members and their children had feared all along. It felt like the whole population had turned against them. It was the moment in which children of collaborators realized that their social position had been dependent on Hitler's occupying forces. For children of collaborators, who experienced the liberation at home, the shock that known and trusted people could turn against them was thus central to their war experience.

The Children Who Fled to the North and East of the Netherlands

Many children of collaborators only realized the dangers they were in when they fled the Allied liberation of the southern provinces in September 1944. Hendrik, the boy who was afraid of anti-NSB signs after liberation, and his mother stayed with family in Amsterdam for a while but returned to their home in The Hague shortly after Dolle Dinsdag. Their final departure came when their house was destroyed in a bombardment in March 1945. At that stage, he left for the rural northeastern province of Drenthe together with his mother and sister. In his memoirs, written in 2006, Hendrik, who as a child had been a big fan of Hitler and a member of the NJS, remembered this journey to Drenthe as a time of adventure. In writing this account, he seems neither to have intended to call attention to "silenced memories" nor to scrutinize his position vis-à-vis his parents and their political choices. This results in a lively manuscript in which, although the child's perspective is presented, innocence or "unfair" treatment are not foregrounded. Hendrik described how he enjoyed walking the roads on their flight to Drenthe and not knowing exactly where their journey would end. During the trip, they called at farms to ask for food and a place to rest. They often encountered hospitality but remained reserved. They presented themselves as bombardment victims, never mentioned their NSB affiliations, and tried hard not to get involved with people. At one of their stops, a local mayor organized a more permanent place for them to stay at a farm near town. Hendrik's family was thus first and foremost seen and treated as bombing victims. This meant that they did not fear liberation as such but more the possible consequences of being "discovered" as collaborators.

When Drenthe was liberated, the stress of hiding their ex-NSB status mounted. As soon as they were unmasked as NSB members, they were separated from the other evacuees. In this memoir it is not the brutal reactions of the guards or local people that loom large. The reactions of locals were overshadowed by the personal bombshell his mother dropped on him at this time, revealing information about his father. She told Hendrik that he had been born while his father was married to another woman and that his father had only joined the SS during the war because he hoped to be able to legally recognize his son. Not only did Hendrik suddenly have to see his father in a different light—he had assumed his father to be a true National Socialist, not one born out of pragmatism—but now, in Hendrik's mind, he himself carried responsibility for their vulnerable position.

At the same time, it is not improbable that there was not much aggression shown by locals to unfamiliar NSB internees. After all, locals had few, if any, personal negative experiences of these Nazi sympathizers, and hence they had less impetus to blame, chastise, or exact vengeance on the strangers. On the contrary, Hendrik also recalled the friends he made when they lived at the farm in Drenthe. One of them, an elderly man, who was also a member of the local resistance, even promised to look after Hendrik's stuffed toy when Hendrik's family was arrested. In contrast, *local* collaborators and NSB members were seen as disloyal to the community. So, unlike the experience of children of collaborators who had stayed at home or whose NSB background was known in their neighborhood, for children like Hendrik the actual liberation period is often not integral to their experience of the war. The flight and their return home with (one of) their parents or being sent to a foster family or children's home had much more impact on their feelings of insecurity. Their idea of living in a hostile world was more connected with the need to keep their "true identities" hidden. Their experiences as "hidden" NSB-ers made their idea of innocence and punishment different from that of the children who had experienced the arrest of their parents at home or whose flight had ended in Germany.

NSB Children in Germany

In September 1944, many NSB families left on special evacuation trains organized by the Germans. The experience of being an expellee and a refugee is central to the accounts of the NSB children who went on this trip. Many of them later recalled that the fact that they had left their homes, their toys, and often their fathers, and faced an uncertain future increased their fear. When these trains came under fire by Allied planes their feelings of vulnerability were amplified. The adults on the journey were unable to offer them protection—in and of itself a frightening and alienating experience—while unknown mighty forces threatened their physical safety. In the memoirs of these refugee children, the image of a burning train in an alien environment (often in an unknown part of the country) came to stand as a metaphor for displacement and isolation in a world filled with enemies.

Memories of trains under fire are common and often depicted in similar ways.[26] Little André was saved by a German soldier, who accompanied his train and threw him over a wire fence. The soldier shot his machine gun at the planes. While his mother tried to protect André by covering him as best she could with her own body, the toddler assured her that when he grew up he too would learn to shoot planes.[27] Some trains managed to keep going, as was remembered by Duke Blaauwendraad-Doorduijn, then an urban middle-class girl of about twelve who had fled with her mother. Her story of a train attack is particularly poignant and harrowing. In the mayhem of the shooting, she saw horrible creatures with mad eyes and striped clothes running in the open field near where she was hiding. Later, she concluded that they must have been psychiatric patients from a nearby hospital.[28] But at the time, she felt as if she was in hell surrounded by shooting, screaming, and unknown mad creatures.

In the stories of these children, the attacks on their evacuation trains are framed as symbolic of their own vulnerability and coincide with a self-image of being refugees and innocent victims. That the trains also often carried German soldiers is subsumed within this framework. The trains under Allied fire are instead remembered as symbols of violence against defenseless women and children and have become separated from the larger historical context.

Duke and her mother were brought to camp Westerbork before going further to Germany. In an ironic twist to her "evacuation" story, it was from Westerbork that the deportation of the Jewish population of the Netherlands had taken place. Almost all Dutch Jews were brought to Westerbork and from there transported to extermination camps, mostly at Auschwitz and Sobibor. Duke, however, did not realize where they were. They met some acquaintances while waiting in a hall near the camp and arranged a private cottage on the campgrounds. That evening they sang songs and played games in order to suppress their feelings of insecurity and disorientation. When Duke went to get food in the morning, she first realized how big the camp actually was. Behind the barbed wire she saw people wearing the Star of David. The last transport of Jews from Westerbork was on 13 September 1944, a few days after the NSB families had left. When her memoirs were published in 1989, Blaauwendraad-Doorduijn did not mention what kind of discussions their presence at Westerbork had evoked among their group. She wrote how she later reflected on having been at a place so tightly connected to the Holocaust:

> After the war when I heard what kind of camp it was I found it terrible that we had been there, there of all places. Only much later did it truly occur to me how horrible this must have been for the Jewish inhabitants [*sic*]. That they were in the anteroom of hell, did not have any future. How unbearable the fears of *those* parents for their children must have been.[29]

Like many others who wrote their memoirs in order to come to terms with their pasts as children of Nazi collaborators, Duke contrasted and compared her experiences with those of other people in danger. Many NSB women and children framed their experiences of flight with the help of existing ideas about people who were persecuted or were on the run. When the NSB families left Westerbork for Germany, they really began to feel like refugees and came to realize that others regarded them that way as well. This is particularly clear in Duke's and Catherine Gosewins's (André's mother's) books.[30] A German woman gave Catherine a piece of bread when they stopped at a German railway station. Accepting the gift, she suddenly realized what she must look like in the eyes of this woman, and she remembered how she had once seen Dutch Jews assembled at the railway station in Amsterdam, waiting with their humble luggage for deportation. One of them had then been eating a rolled up pancake and she had wondered how someone in such a state of degradation could possibly think of eating.[31] This memory exemplifies how the experience of being a refugee simultaneously collided and merged with earlier ideas about people on the move. The realization that now it might be *their* turn to be expelled, degraded, and dependent on the help of others came as a shock. Implicitly, Catherine shows in her description of this moment that while she felt expelled and vulnerable like other "expellees," she also did not confront herself with the uncomfortable questions about why she had not been interested in the deportation of the Jews or whether the comparison she made was fair and not merely a way of assigning the status of victimhood to herself.

The awareness of their new identities as refugees was further enhanced when NSB families arrived at their destination in Germany. Duke remembers how an arrogant Dutch NSB official awaited them and directed them through the village where people stopped to stare at them.[32] The locals were already getting used to the fact that large groups of foreigners were brought to their town. In the last stages of the war, millions of people from abroad stayed as refugees, POWs (prisoners of war), or as compulsory, forced, or slave laborers in Germany. Often locals did not know whether they had friends or foes among them. They were not necessarily friendly to Dutch Nazi sympathizers. Sometimes the Dutch were, just as in the Netherlands, seen as traitors to their own country, at other times as annoying extra mouths to feed, as Kitty, who was then seven years old, remembered.[33] Yet Rinnes Rijke, who was brought to Germany by his father, remembers it differently: in the town where he arrived, many people were friendly and gave him sweets or fruit.[34]

While not necessarily welcomed with open arms, the Dutch National Socialists in Germany enjoyed a position that was different than the one they had occupied in the Netherlands. In Germany, they were no longer confronted with a large majority that regarded them as traitors. Their newly constructed identities as refugees could thus be further reinforced. This is particularly important for the children of collaborators: in describing their stay in Germany, they do not mention any fear of, for instance, being "discovered" or punished but focus on ordinary hardships for (child) refugees during war, such as bad housing, lack of food, and the fear of bombings. Thus, their accounts may resemble those of German "children of the war." But sometimes they also stress the differences between themselves and the Germans, who as a result of worsening conditions in the evacuation camps became more and more the enemy for some.[35] Consequently, the self-image of being a victim of the Germans could also take root.

In Germany, it soon became clear that the officials of the Nationalsozialistische Volkswohlfahrt[36] regarded Dutch Nazi supporters as welcome extra support for the German war effort. Apart from daily work on farms and in factories for the adults, many of the older children were (often consensually) sent elsewhere for "war service." An unknown number of children were separated from their families. Iet, then a girl in her early teens, was disappointed when her sister was sent to a camp in the Sudetenland and she was not. The 2 sisters and their mother expected that this camp was something of a boarding school where she would receive a good education and have her own room. This turned out to be a major disappointment, and the letters sent from the Sudetenland show regret, strong feelings of homesickness, and, at the same time, self-reassurances that the 12-year-old would pull through, particularly since she knew that the soldiers at the front had a much heavier burden to bear.[37] Neither of the sisters ended up in the German war industry although this happened to many girls who were sent to the eastern parts of Germany. Many boys were taken to Wehrsportlager (army sports camps) in Austria where they received military training. Some 400 to 500 Dutch boys thus ended up in the Waffen SS.[38] Parental reactions to their sons' military training were varied. Some mothers were particularly panicked about their children being trained for frontline service. Others fully supported the fact that their children were fulfilling their duty for the German Reich.

The children of collaborators directly experienced war when the German cities near their camps were bombed or when they became combatants and actual participants in the German war machine. When the latter is described in a memoir, it is usually depicted as an unwilling consequence of German policy and as something horrible to happen to a child. Parents' responsibility for this participation in the war effort or the consequences of the children's own choices are seldom discussed in the memoirs. The framework of the stories is that of the innocent child who is swallowed up by the war. A telling example is the story of Rinnes Rijke. One day in the autumn of 1944, Rinnes was taken to a Hitler youth house (HJ-Heim) in Hanover. Rinnes was eleven years old and, in fact, too young for the HJ. Although he felt quite proud that he had a uniform, he was also the smallest boy in the home, the only one from the Netherlands, and totally unaccustomed to the militarized way of life in an HJ-Heim. His father only visited him once and, instead of comforting Rinnes and taking him away, as Rinnes had hoped he would do, the man broke down and talked about suicide. Rinnes's father returned to the Netherlands, where he worked for the Germans and remarried. Rinnes desperately awaited letters from his father, but when none arrived, he got used to getting by on his own and trusted that all would be well once the war was over and he could go home. Rinnes was soon sent to an HJ-Heim where he was to have his military training and where the discipline was harsh and humiliating physical punishments often occurred.[39]

Rinnes's memoirs were published in 1982. His book was one of the first published accounts of a collaborator's child and, as such, received considerable attention. Rinnes framed his memories around the title *Niet de schuld, wel de straf* (*Not Guilty, but Still Punished*). In his book, he focused on the revenge that was exacted on him by the Dutch public for being the son of a collaborator. Rinnes did not analyze what had happened to him since the framework of "punished without guilt" made that superfluous. He was not looking for ways to understand his past or the relationship with his family and society but was rather seeking to convince his readers of the unjust treatment he received after the war in the Netherlands.[40] Interestingly, he did not present his period of "exile" in Germany in 1944 and 1945 in terms of injustice and punishment. Conditions at the HJ house were harsh, but he accepted it as a part of what war was about.

The return to the Netherlands, especially for children who had lost contact with their parents, could be a troubled trip. After the German surrender, millions of displaced persons had to be repatriated. Among them were about 300,000 people from the Netherlands, including 270,000 Dutch workers (mainly men) who had been forced to work in Germany, about 13,000 Dutch men and women who were released from concentration camps, and 10,500 Dutch POWs.[41] According to official sources, there were about 4,500 Dutch willingly residing in Germany at the end of the war, including volunteers in the German military forces and collaborators, many of whom were Dolle Dinsdag refugees.[42] It is unclear whether these statistics included their children. It is possible that in the chaos of the May 1945 period, children traveling alone or in small groups did not stand out. Lia was with other NSB children in Theresienstadt at the end of the war and remembered that liberated Jewish women took care of them. In mentioning this, she also referred to the maltreatment she received on her return to the Netherlands. Tom, then aged 12, and Nico, 10, had similar experiences on returning from a Kinderlandverschickung (a holiday camp in the countryside, KLV) in Czechoslovakia.[43] Ida also traveled back through the frontlines. She was questioned by Dutch border guards, and it was at that stage, as she explained it, that her humiliation began. Henk, who was then 12, also remembered this.[44] As soon as the children encountered other Netherlanders, the enormous cleavage between the collaborators and their children and the rest of Dutch society became apparent. The children feared that Bijltjesdag was imminent. "We will hang you all from the highest trees," warned a man Rinnes met on a train to Belgium.[45]

Rinnes himself did not encounter any bad treatment on his return to the Netherlands, and even though many Dutch officials did not seem very friendly, there was always someone willing to take care of him. Rinnes lied about his past, keeping silent on his NSB and HJ connections and claimed that he had been sent to a children's holiday camp in Germany. The border administration dismissed him and sent him home. The real disappointment occurred when he found out that his father had married a woman who instantly began taking out all her frustrations on him. Rinnes urged his readers to share his opinion that that it was "Dutch society" that had let him down, assigning him the stigma of being a war criminal and failing to rescue him from his evil stepmother. Rinnes's maltreatment at the hands of his stepmother symbolizes for him what society did to him: where he had longed for a safe home, instead he found ongoing hardship. It is noteworthy that Rinnes avoids examining the relationship with his father, his mother, and his stepmother and instead focuses on the punishment inflicted on children of collaborators by society at large. He appealed to the readers to stand up and exclaim that such a society should be ashamed of itself and that surely they did not want to belong to a community that punishes innocent children.

An example of an autobiography that revolves totally around the topic of unjust punishment by society was written by "P. Berserk."[46] "Berserk" was one of the few Dutch boys from a working-class background who attended a German cadet school. His father and elder brothers all fought on the eastern front. "Berserk" was bullied at school and was, therefore, sent to a German school, which did not make things any easier since he had to wear his uniform publicly, making him an easy target for Dutch children in the streets. In autumn 1944, he and his schoolmates were sent to build trenches on the western front. They had to work on evacuated land in the Belgian-German border region. They received meager rations and the German boys blamed everything that went wrong on the Dutch boys. When the fighting reached their positions, "Berserk" was already suffering from hunger and a lice infestation. In the midst of the raging of the war, while being shot at by American fighters and listening to the German anti-aircraft guns responding, it dawned on him that the whole *Übermenschen* story was one big lie. His belief in National Socialism further crumbled when a group of Red Cross nurses came to powder the naked bodies of all men present against scabies. Like "white Negroes," "Berserk" wrote, we were dancing and screaming in the night.[47] After that they were sent to the German heartland. "Berserk" ended up near the border with Poland at the end of 1944 and decided to look for his father, who was supposed to be in Posen at that time. He soon ended up going west again with the refugee masses fleeing from the Russian troops. Then all hell really broke loose. His impressions of this period consisted mostly of lying in the mud in the burning cities of Dresden, Berlin, and Hamburg with his hands pressed against his ears against the thunder of bombardments.[48] He managed to get to Schleswig Holstein, where he worked for various farmers until Dutch officials discovered him at the end of the summer and returned him to the Netherlands to be put on trial for serving in enemy forces.

In contrast to Rinnes's reception in the Netherlands, in cases like "Berserk's" it was obvious that youngsters returning to the country had participated in the Nazi war effort. The stress on innocence in these memoirs is in direct opposition to the way these adolescents were often seen at the time: as Hitler's soldiers. "Berserk" may have been a child soldier, but he was still seen as being active in the German war effort and was consequently regarded as a danger to the Dutch and Allied cause and as a person deserving punishment for his pro-Nazi acts. When "Berserk" returned to the Netherlands, for example, he was beaten and humiliated by border guards.[49]

On their arrival in the Netherlands, the expectation many children had of Bijltjesdag became reality. Anticipated and experienced events mingled with rumors and stories that filled the border and internment camps. Central to all the stories of repatriation is the uncertainty about their own position and future. At the time, women were rumored to have been raped, men beaten and shot at, children taken away, maltreated, and put in detention centers until they reached the age of maturity. Many camp guards were indeed lax about keeping to formal prison rules. They were volunteers who lacked a professional background and often felt that they were guarding the conquered enemy giving them the right to abuse the internees. Camp commanders did not always discipline their guards for the abuse of the detainees and sometimes they even encouraged it.[50] "Berserk" described how he was forced by the guards to jump around like a frog, a humiliation often mentioned by interned collaborators,[51] but that he refused:

I'd rather die than do *that*, it runs through my head. They lock me in a dark room. I am hungry. I have wet my trousers. My face bleeds. I am in utter darkness and weep like a child. *Those* are the images of my adolescence! Your liberation was my mental death![52]

This last exclamation illustrates how "Berserk," like Rinnes, framed his story as an accusation and wanted to invoke in Dutch readers compassion and feelings of shame that their society allowed these things to happen to innocent children. "Berserk" most probably did not choose his pseudonym by accident. His memoirs, published shortly after Rinnes', were fragmented and interwoven with outcries about him getting angry and violent while writing down his memories. He raged against the society that had made this human wreck out of him.

Duke is much more inquisitive than both Rinnes and "Berserk." She and her mother returned home via Bremen, where Dutch repatriation officials interrogated them. When Duke refused to give up the names of other NSB families in her street, one of the men grabbed her, and she suggests in her text that he sexually assaulted her. She avoided specifying what happened and appeared to shy away from putting this memory explicitly on paper.[53]

To her this experience may have been an example of her ultimate humiliation not just as a child of collaborators but also as a girl. She avoided telling her mother about what had happened, fearful that her mother's reaction would further endanger their position. Later she witnessed the humiliation of other collaborators. When they arrived back in the Netherlands, Duke wrote that they had to watch how men had to run circles in a schoolyard while being shot at.[54] At the border they were, like all repatriates, disinfected. Duke described this as something she found very humiliating. Having DDT sprayed on her and the other repatriates' naked bodies symbolized their vulnerability and invoked associations with vermin that needed to be destroyed.[55] The act of disinfection signified to her that she would, from now on, be treated as a second-class citizen by her own people and became, therefore, a part of the memorial framework of social punishment.

For "Berserk" and Rinnes it was not so much the failure of their families to prevent their participation in the German war effort or even that some parents actually sent their children off to participate in the war that was central to their narratives. Rather, it was that these experiences at the front were seen in the Netherlands as proof of their own, personal, guilt and as evidence that they were traitors to the national cause. Society labeled them "war criminals," while they identified themselves largely as victims. The source of this victimization was thus, in their minds, transferred from the war itself to the social reaction to it afterwards. For many collaborators' children who returned from Germany, the public's condemnation of these young people's actions became a defining experience. The shock of being regarded as perpetrators of war crimes dominated their memories, exacerbating the image they had of themselves as damaged children who were innocent victims of the war.

Conclusion

Not only were collaborators' children's experiences of war far from uniform, the ways in which their memories were framed also varied. Some stayed in the Netherlands and experienced the liberation of their homeland and subsequent arrest of (one of) their parents at home. These arrests were often accompanied by violence. A central experiential element of the war for these children was the loss of trust in their familiar environment: neighbors, friends, and acquaintances could become dangerous people who had the right to arrest and humiliate their parents and to hit or bully the children. Other children were billeted with farmers in the eastern or northern provinces in the last stages of the war and were identified after liberation not as evacuees but as "political delinquents." The possession of a "secret identity" that must be kept hidden is one of the central elements in their war experience. Feelings of loss often manifested themselves later when some of these children were sent to foster families or children's homes. This too became remembered as unjust punishment in later years.[56]

Children who went to Germany often experienced fighting and bombardments. Their flight, self-identification as refugees, and sometimes personal experience of fighting at the front created identities of victimhood that were not accepted when they returned to the Netherlands. The disillusionment they experienced when their suffering was not acknowledged and, worse, when they were treated as "collaborators" is a significant element in their memories. The long-hoped-for return home was a disappointment. "Berserk" and Rinnes present this as a betrayal by Dutch society: instead of upholding the ideal of the innocent child and complying with his or her need for help, safety, and love, society instead betrayed the child that returned home from a nightmare. Instead of safety and normality, the end of the war seemed only to usher in a new phase of continued suffering, humiliation, and rejection.

The reconstruction of liberation and internment experiences in the memoirs has been influenced by the belief that postwar society punished not only Nazi collaborators but also their "innocent" children. Since the late 1970s, it has become increasingly common to show compassion for the former children of collaborators and to encourage them to tell their stories within the framework of being unjustly punished for the actions of their parents. This framework fits in with more critical Dutch debates about the Second World War with regard to national myths of heroism and resistance that began in the late 1960s. Consequently, the children of collaborators found ways to tell their stories that in the social contexts of the 1940s, 1950s, and 1960s had been ignored or met with disapproval. In the immediate postwar years, the crimes of their parents, rather than the impact of the war on these children, had been the focus of public comment and criticism.

The primary narrative construction of the lives of the children of collaborators as histories of social retribution in the postwar period has obscured their experiences of the war itself. The focus on the ways in which these children were "punished" by society has also overshadowed the role that their parents played in their wartime experiences. Some authors have scrutinized their childhood memories and tried to understand their relationship with their parents and the effects of their parents' choices. Duke Blaauwendraad-Doorduijn and Sytze van der Zee, for instance, undertook such a journey and, thus, were able to provide a much more multifaceted impression of the repercussions of the Second World War than the accounts of Rinnes and "Berserk." Regardless of the kind of memoir produced, however, what remains is a legacy of children whose lives were irrevocably changed and damaged by war. Whether victims or naive participants, the children of Nazi collaborators demonstrate, through their stories how the use of the "innocent child" label enabled Dutch public discourse to address the topic of collaboration and slowly integrate it into the broader Dutch collective memory of the war. These memories are of great importance when studying the legacy of collaboration. The specific narrative of unjust punishment by society is, however, still a blinkered one when it comes to questions of innocence, responsibility, and guilt. Furthermore, experiences that do not fit into this framework remain as silent as the children themselves, their voices still drowned by the cacophony of "acceptable" war stories.

Notes

1. I. P. Spruit, ed., *Onder de vleugels van de adelaar. Kind van de Führer. Levensverhaal van een Nederlandse ex-SS'er, opgetekend door Inge P. Spruit.* Bussum, The Netherlands, 1983.

2. In the days following *Dolle Dinsdag*, approximately 65,000 collaborators and their families fled to the north of the Netherlands and to Germany: L. de Jong, *Het Koninkrijk der Nederlanden in de Tweede Wereldoorlog, 10b Eerste helft: Het laatste jaar, deel II.* The Hague, 1980, p. 281.

3. J. C. H. Blom, "Lijden als waarschuwing. Oorlogsverleden in Nederland," *Ons Erfdeel.* 4, 1995, pp. 531–541.

4. J. Vanderwal Taylor, *A Family Occupation. Children of the War and the Memory of World War II in Dutch Literature of the 1980s.* Amsterdam, 1997, p. 19.

5. For example: Stichting Icodo, *Oorlogskinderen: toen en nu.* Utrecht, The Netherlands, 1995; F. A. Falch, *Kinderen van . . . . : interviews met de naoorlogse generatie.* Leiden, 1999; E. E. Werner, *Through the Eyes of Innocents: Children Witness World War II.* Boulder, CO, 2000; C. Landgraf and R. Pfirschke, eds., *Unterwegs mit Koffer und Teddybär: Europas Kinder und der Zweite Weltkrieg.* Riga, Latvia, 2005.

6. Sources for the memories and experiences of these children include press clippings ("Knipselarchief," Netherlands Institute of War Documentation [NIOD], Amsterdam); B. Kromhout, *Fout geboren. Het verhaal van kinderen van foute ouders.* Amsterdam; Antwerpen, Belgium, 2004; M. Lindt, *Als je wortels taboe zijn. Verwerking van levensproblemen bij kinderen van Nederlandse nationaal-socialisten.* Kampen, The Netherlands, 1993; G. Scheffel-Baars and P. Mantel, "NSB-kinderen in tehuizen." Unpublished manuscript, 1987; T. Vorst-Thijssen

and N. de Boer, *Daar praat je niet over! Kinderen van foute ouders en de hulpver-lening*. Utrecht, The Netherlands, 1995 (1993); various published and unpublished autobiographies mentioned in the references below. This article is part of a broader research project on children of Dutch collaborators in early postwar society (1945–1960) that includes the study of literature, archival sources, personal documents, and interviews. Publication of a monograph on this subject is projected for autumn 2008.

7. See J. C. H. Blom, *In de ban van goed en fout?: wetenschappelijke geschiedschrijving over de bezettingstijd in Nederland*. Bergen, Norway, 1983; A. D. Belinfante, *In plaats van bijltjesdag: de geschiedenis van de bijzondere rechtspleging na de Tweede Wereldoorlog*. Assen, The Netherlands, 1978; I. de Haan, *Na de ondergang: de herinnering aan de jodenvervolging in Nederland, 1945–1995*. The Hague, 1997; J. Withuis, *Erkenning : van oorlogstrauma naar klaagcultuur*. Amsterdam, 2002.

8. Vanderwal Taylor, *Family Occupation*, p. 16.

9. Ibid., p. 19.

10. The NSB (*Nationaal Socialistische Beweging*) was the Dutch Nazi party, founded in the early 1930s.

11. This remark is often made in interviews. See also Armando and H. Sleutelaar, *De SS'ers: Nederlandse vrijwilligers in de Tweede Wereldoorlog*. Amsterdam, 1967, p. 435; R. Rijke (Piet van Weelden), *Niet de schuld, wel de straf. Herinneringen van een NSBkind*. Bussum, The Netherlands, 1983(1982), p. 167.

12. "De eerste tien jaar van de oorlog 1940–'45. Van een oorlog die maar niet voorbij wil gaan." Unpublished manuscript, 2006. The author stated in a personal communication that he desires to remain anonymous.

13. For example, M. Diederichs, *Wie geschoren wordt moet stilzitten. De omgang van Nederlandse meisjes met Duitse militairen*. Amsterdam, 2006. The practice of shaving women's heads was not unique to the Netherlands and also occurred in France, Belgium, and Norway: F. Virgili, *La France "virile": des femmes tondues à la libération*. Paris, 2000; E. B. Drolshagen, ed., *Nicht ungeschoren davon gekommen: das Schicksal der Frauen in den besetzten Ländern, die Wehrmachtssoldaten liebten*. Hamburg, Germany, 1998.

14. S. van der Zee, *Potgieterlaan 7. Een herinnering*. Amsterdam, 1997, p. 23.

15. Ibid., p. 43.

16. Ibid.

17. Ibid., p. 176.

18. Ibid., p. 58.

19. Scheffel-Baars and Mantel, "NSB-kinderen in tehuizen," p. 61. The names of the interviewees were altered in the thesis.

20. Kromhout, *Fout geboren*, pp. 62–63. The names of the interviewees were altered in the book.

21. Ibid., p. 63.

22. Ibid., pp. 62–63.

23. Scheffel-Baars and Mantel, "NSB-kinderen in tehuizen"; Kromhout, *Fout geboren*.

24. For instance, I. van Bekkum, "Vlucht naar Duitsland." Unpublished manuscript, 2003; C. Gosewins, *Een licht geval.* Amsterdam, 1980, p. 165; Scheffel-Baars and Mantel, "NSB-kinderen in tehuizen"; Kromhout, *Fout geboren.*

25. de Jong, *Het Koninkrijk der Nederlanden in de Tweede Wereldoorlog, Deel 12.* The Hague, 1988, p. 498.

26. See also the stories of "Miep" and "Jan" in Kromhout, *Fout geboren*, pp. 43, 46.

27. Gosewins, *Een licht geval,* p. 127.

28. D. Blaauwendraad-Doorduijn, *Niemandsland.* Amsterdam, 1989, p. 33.

29. Ibid., p. 35.

30. Also "Mijn verhaal had niet verteld mogen worden. Een zwijgende generatie sterft uit." Unpublished manuscript, 2006. The author stated in a personal communication that she desires to remain anonymous.

31. Gosewins, *Een licht geval,* p. 129.

32. Blaauwendraad-Doorduijn, *Niemandsland,* p. 39.

33. Scheffel-Baars and Mantel, "NSB-kinderen in tehuizen," p. 59. Also: Vorst-Thijssen, de Boer; Kromhout, *Fout geboren.*

34. Rijke, *Niet de schuld, wel de straf,* pp. 30, 39–40.

35. For instance: "Mijn verhaal had niet verteld mogen worden."

36. *Nationalsozialistische Volkswohlfahrt* (NSV) was the Nazi welfare organization that took care of German refugees from Eastern Europe and fleeing collaborators from the West.

37. Van Bekkum, "Vlucht naar Duitsland," p. 26.

38. N. C. K. in't Veld, *De S.S. en Nederland: documenten uit S.S.-archieven 1935–1945.* The Hague, 1976; de Jong, *10b*, p. 291.

39. Rijke, *Niet de schuld, wel de straf,* p. 93. Also: Spruit, *Onder de vleugels van de adelaar.*

40. J. Vanderwal Taylor, "Rinnes Rijke's Niet de schuld, wel de straf as a Social Phenomenon: An Attempt to Come to Terms with a Tragic Past," *Canadian Journal of Netherlandic Studies.* 2, 12, 1991, pp. 28–32.

41. I. de Haan and J. W. Duyvendak, *In het hart van de verzorgingsstaat. Het Ministerie van Maatschappelijk Werk en zijn opvolgers (CRM, WVC, VWS), 1952–2002.* Zutphen, The Netherlands, 2002.

42. M. Bossenbroek, *De Meelstreep.* Amsterdam, 2001.

43. Scheffel-Baars and Martel, "NSB-kinderen in tehuizen," pp. 58–60.

44. Ibid.

45. Rijke, *Niet de schuld, wel de straf,* p. 167.

46. P. Berserk [pseud], *De tweede generatie: herinneringen van een N. S. B.-kind.* Utrecht, The Netherlands, 1985.

47. Ibid., p. 60.

48. Ibid., p. 62.

49. Ibid., pp. 66–68.

50. Despite various public debates and a parliamentary enquiry in the late 1940s, the general feeling in the Netherlands seemed to have remained that what had happened in the internment camps of ex-collaborators was unsurprising given

the harsh occupation regime the Dutch had suffered: de Jong, *10b*; A. D. Belinfante, *In plaats van Bijltjesdag: de geschiedenis van de Bijzondere Rechtspleging na de Tweede Wereldoorlog*. Assen, 1978; P. Romijn, *Snel, streng en rechtvaardig: politiek beleid inzake de bestraffing en reclassering van "foute" Nederlanders, 1945–1955*. Houten, The Netherlands, 1989.

51. It seems to have been a common form of humiliation; many who served in the SS also remember it from their time in the SS-instruction camps.

52. Berserk, *De tweede generatie*, p. 68.

53. Blaauwendraad-Doorduijn, *Niemandsland*, p. 89.

54. Ibid., pp. 91–92.

55. For other examples, H. Piersma, ed., *Mensenheugenis. Terugkeer en opvang na de Tweede Wereldoorlog*. Amsterdam, 2001.

56. I. M. Tames, "Children of the 'Enemy Within'. Reintegration of the Children of Dutch Collaborators in the Early Postwar Years." Unpublished paper presented at the SHCY Conference "In the name of the child." Norrköping, Sweden, June 2007.

CHAPTER 5

The Child Soldier in Literature or How Johnny Tremain Became Johnny Mad Dog

David M. Rosen

Nearly every day, a world-weary battle-hardened child stares out at us from a newspaper or magazine. Every time we turn on the television or radio, we are confronted with another story of children bearing arms. In all of the rebellions, insurgencies, and civil conflicts that now involve millions of people across the globe, there is one common and undisputable fact: children and youth are always on or near the field of battle. This brutal truth is terribly unsettling. Indeed, the image of child soldiers has become a powerful symbol of nearly everything that is wrong with war.

The prevailing view is that child soldiers are the victims of adult abuse and criminality. They exist as the most transgressive form of noncombatant: children who have been forcefully and unlawfully transformed into combatants in violation of their essential qualities. Like the concepts of child laborer, child bride, or child prostitute, the child soldier is seen to be a deviant product of adult abuse, and the presupposition is that these children are dependent, exploited, and powerless. Even where a child may have committed terrible war crimes, the child's culpability is attributed to adult misuse and exploitation.

Our current understanding about child soldiers has been primarily shaped by an emerging international humanitarian discourse about children. Found primarily in the reports of nongovernmental organizations, such as Human Rights Watch, Amnesty International, the Coalition to Stop the Use of Child Soldiers, and numerous others, this discourse has had a profound effect upon public consciousness. But this discourse evidences little or no awareness that current humanitarian views about childhood derive from a particular constellation of ideas and practices that began to emerge in Europe at the end of the Middle Ages. During the Middle Ages, children were deemed to be the natural companions of adults.[1] But it was also a time when a new set of new ideas about childhood developed that stressed belief in the innocence of childhood, the practice of segregating children from adults, and the isolation and prolongation of childhood as a special protected state. These ideas and practices were virtually unknown in the preindustrial world but developed and spread in the West with the industrial revolution, until they were established, albeit unevenly, across virtually all class and cultural boundaries.

The emergence of formal and institutionalized schooling during the industrial revolution served to strengthen the idea of the innocence and weakness of children and to increasingly segregate young people from adults. Adolescence, it has been quipped, was invented with the steam engine.[2] During the industrial revolution, schooling slowly replaced apprenticeship as the prime mode of education. Traditionally, military training was tied to the apprenticeship system and was the most resistant to formal schooling of all the professions. In the seventeenth century, a boy destined for a career in the military—the so-called noble profession—would have perhaps two or three years of separate education and at the age of 11, 12, or 13 would find himself as a commissioned officer in the army or navy, freely mixing with adults in the military camps.[3] Historically, soldiering appears to be one of many professions that by necessity ignored the growing separation of children from adults—where else, after all, were the next generation of recruits to come from?

But even schooling itself and its associated ideas of childhood were not necessarily incompatible with military ideals. As schooling began to dominate educational processes, there was a simultaneous union of military and school cultures, as schools, which had once been primarily ecclesiastical institutions, became militarized. So, as formal education began to separate child life from adult life and create a special culture of childhood, that culture itself was shaped by a military ethos. Military discipline was deemed to have a particular kind of moral virtue. To the extent that military life was understood to be virtuous and ennobling, there was little conflict between the idea of the child and the life of the soldier.[4] By the end of the eighteenth century, the formal relationship between children and military life was frequently organized through a variety of institutional mechanisms that combined military training, apprenticeship, and pedagogy in varying combinations according to class and status. This pattern continued well into the middle of the twentieth century.[5]

Humanitarian discourse has had an equally profound effect upon contemporary literary conventions and has reversed the images of children under arms that pervaded much of nineteenth- and twentieth-century literature. The transformation is startling: the heroic child fighters of yesteryear, such as Gavroche in Victor Hugo's *Les Miserables* or the boy spy, Kim, in Kipling's eponymous novel, have been replaced by Agu, the battered victim of a nameless war in Uzodinma Iweala's *Beasts of No Nation*. This is not to suggest that there were no dissenting views in the past. As early as 1861, Herman Melville raised his skeptical voice against the chorus of hosannas surrounding young boys marching off to the civil war. In his poem *The March into Virginia Ending in the First Manassas*, he writes, "All wars are boyish, and are fought by boys, the champions and enthusiasts of the state."[6] But Melville was in the minority, and it took more than 100 years before his lone voice became part of the cacophony of humanitarian discourse.

How did the heroic child soldier of an earlier era come to be replaced by the abused and exploited child who is both killer and victim? What alterations in literary conventions and moral attitudes were required in order to transform the child soldier into its modern literary construction? Much of this stems from the intense focus on African conflicts. While children have been recruited as child soldiers in wars all over the world— Columbia, Kurdistan, Laos, Mexico, New Guinea, Pakistan, Palestine, Peru, the Philippines, Sri Lanka, and New Guinea come immediately to mind— the contemporary literary gaze remains firmly fixed on Africa. Exactly why this is the case is unclear. Certainly some contemporary examples of the use of child soldiers in Africa, such as the Revolutionary United Front in Sierra Leone and the Lord's Resistance Army in Uganda, have provided chilling examples of the abuse of children. But these extraordinary cases have also come to serve as the archetype of child soldiers' experiences in both Africa and elsewhere. Literary treatments of African children at war, almost all geared to Western audiences, magnify this perspective by the lingering tendency to see Africa with Conradian eyes: seeing in Africa only "the heart of darkness." The general Western discourse about war in Africa, whether precolonial, colonial, or postcolonial, has remained remarkably consistent since the middle of the nineteenth century. In this discourse, warfare in Africa—in contrast to warfare in the West—is invariably cast as irrational and meaningless.[7]

Our understanding of war has also been affected by a more than half a century of peace (with obvious exceptions) in the West. Accordingly and luckily, we have lost a visceral understanding of war. Instead, our experience of war is mediated by cultural and geographical distance, professional volunteer armies, civil society, and human rights organizations, all of which, in myriad ways, serve to ascribe war to an essentialized "other." As distant observers, we remain the ultimate noncombatants with little knowledge of the kind of warfare at home that often thrusts children into combat. From the safety of the West, we may have reached a point where we can barely comprehend the agility and resourcefulness of the children Anna Freud encountered during the years of the Nazi air blitz in London.[8]

The Child Soldier in Modern Humanitarian Discourse

The concept of the "child soldier" seems to be the conflation of two contradictory and incompatible terms. The first, "child," typically refers to a young person between infancy and youth and connotes immaturity, simplicity, and an absence of full physical, mental, or emotional development. The second, "soldier," in the context of contemporary professional armies in the West, generally refers to men and women who are skilled warriors. Indeed, the entire concept of the child soldier melds together two very contradictory and powerful ideas, namely, the "innocence" of childhood and the "evil" of warfare. Thus, from the outset, in modern Western imagination, the very idea of the child soldier seems both aberrant and abhorrent.[9]

Distaste for the idea of the child soldier is most clearly expressed in current attempts by humanitarian groups to create an international ban on the recruitment of child soldiers. Most of these groups have adopted the so-called Straight 18 position that defines childhood as beginning at birth and ending at age 18 and defines the child soldier as any person under 18 years of age who is recruited or used by an army or armed group.[10] Humanitarian groups have succeeded in altering the military recruiting policies of many countries and, perhaps most importantly, in bringing about changes in international humanitarian law, the so-called laws of war, so as to make the recruitment of children below the age of 15 a crime.

Part of the problem is the very novelty of the modern concept of the child soldier. The "Straight 18" position is a prime example of how a new political agenda can be represented as an existing cultural norm. It mandates an international acceptance of two major principles: first, that childhood be universally defined as beginning at birth and ending at age 18, and second, that childhood is incompatible with military experience. At its heart it requires that other existing and competing definitions of childhood be abandoned in favor of a single international standard. Cast in the language of human rights and humanitarian imperatives, this definition of childhood pays little attention to the enormity of the issues of social and cultural changes contained in the transnational restructuring of age categories. Like many other avowed human rights imperatives, it tends to ignore or demonize the historical experiences and moral and legal imperatives of other cultures.[11] Moreover, in adopting a single universal definition of childhood, both international humanitarian organizations and human rights law ignore that there is no universal experience and understanding of childhood. Indeed, if literature has anything to contribute to an understanding of childhood, it is that it gives voice to a multiplicity of childhoods, each culturally codified and defined by age, ethnicity, gender, history, and location.

Humanitarianism and literature tend to narrate the experience of childhood in antithetical ways. Humanitarian law codifies bright-line distinctions between childhood and adulthood that are largely indifferent to context; literature understands context as informing virtually all narratives about children, including distinctions of age. Humanitarian discourse on the victimization of children also contrasts with literary conventions that set children into roles as active players and participants in society. Most modern literary forms, including the novel, force a focus on individuals and their engagement with surrounding psychological or social forces. Indeed, the development of modern literature and the development of character go hand in hand. By way of contrast, humanitarian rhetoric and reporting about child soldiers work against character as they strive to create essentialist categories with universal applicability. As a result, humanitarian narratives tend to be breathtakingly superficial and thin and bear scant relationship to the experience of children at war traditionally found in literature, anthropology, or history.

Anthropologists have long been aware that there is no fixed single chronological age at which young people move from childhood to adulthood and enter into the actions, dramas, and rituals of war.[12] Warfare draws in the young and the strong. The transition to warrior probably turns on a wide variety of practical issues since young people, mostly boys, would have to be in a position to personally demonstrate their physical and emotional fitness for these roles. The overall picture suggests that chronological boundaries between childhood, youth, and adulthood are highly varied and rooted into the historical experience of each society and culture. Indeed, it is hardly clear whether all societies even make use of these or similar concepts of childhood.

Similar issues arise in Western societies. Until recently, the armies of Western Europe and the United States were filled with "boy soldiers." Boy soldiers have been routinely recruited into the British military since the Middle Ages, and by the late nineteenth century, various institutions emerged that organized and systematized their recruitment.[13] A wide variety of data also indicate the presence of the very young on the American side of the Revolutionary War.[14] Until the twentieth century, most military service in the West was voluntary, but even with the emergence of conscription, the recruitment of child soldiers continued as schools and military apprenticeship programs continued to channel boys into the military.

The Civil War in the United States was a war of boy soldiers. Throughout the Civil War, youngsters followed brothers, fathers, and teachers into war. They often had support roles, but quickly graduated into combat roles. They were sometimes recruited at school and, when necessary, used weapons that were cut down and adapted for use by younger people. Numerous examples of famous boy soldiers abound: David Baily Freemen, "Little Dave," enlisted in the Confederate army at age 11, first accompanying his older brother as an aide-de-camp and then as a "marker" for a survey team, before finally fighting against Sherman's army.[15] Avery Brown enlisted at the age of eight years, 11 months, and 13 days in the Ohio Volunteer Infantry. Known as the "Drummer Boy of the Cumberland," he lied about his age on his enlistment papers, listing it as 12.[16]

Of equal importance is how the participation of boy soldiers in war was understood. Writings about boy soldiers in the aftermath of the Civil War constitute a hagiographic genre celebrating the nobility and sacrifice of young boys in battle. The existence of developmental differences between boys and men were recognized in this literature but understood rather differently than they would be today. Although young boys were regarded as impulsive and less mature than older men, these qualities were recast as grand and heroic. Testimonials collected by Susan Hull in 1905 describe boy soldiers as enduring battle with "patience and gaiety" and those who died as having "made their peace with God." Equally important, the experience of battle, however horrific, was not understood as destroying the lives of children but as ennobling them. Boy soldiers who survived intact were described as respected citizens whose contribution to civic life was enhanced by their experience of war.[17] While it may not be possible to verify the accuracy of these accounts, they are conspicuous precisely because they put forward radically different views of children in battle than those contained in contemporary humanitarian accounts.

There is no doubt that hagiographic accounts also mask the brutality to which young people are (and were) exposed during war. Nevertheless, these not-so-distant descriptions of boy soldiers make it apparent that current humanitarian views of children at war are very different from the way this was understood in America and Europe in earlier times. The current view of the child soldier, as an abused and exploited innocent, is a radically new concept that is linked to the deeply entrenched but equally modern view of the child as pure and unspoiled and as the ultimate victim of war.

The Child Soldier in Literature

The transformation of the child soldier, from hero to killer or victim, is equally vivid in literature. The classic nineteenth-century representation of the child at war is the character of the street urchin, Gavroche, in Victor Hugo's *Les Miserables*. The character of Gavroche is at least partly based on an existing icon of the child soldier found in Eugene Delacroix's nineteenth-century painting *Liberty Leading the People*.[18] The painting depicts a scene at the barricades during the July Revolution, 1830. At the center of the painting is "Liberty" in the form of a woman leading the charge over the barricades while clasping the flag of the French revolution in one hand and a musket in the other. To her immediate left is an equally powerful portrait of a child, a young boy, brandishing a musket in each hand. The child under arms was often thought to serve symbolically as a personification of class struggle. Armed children represented the lofty goals of popular insurrection that drew people from all walks of life into the battle against monarchy and entrenched privilege.[19]

Les Miserables was written some 32 years after these events. A key moment in Hugo's novel is the Paris student uprising of June 1832, where many of those who die are students involved in a short but violent antimonarchist revolt. As with Delacroix, the main action is on the barricades and focuses on a child, the orphan Gavroche, a street urchin who joins with the student rebels, pistol in hand. During the battle, he crosses over the barricades into the line of fire in order to gather unspent cartridges from among the dead. He is killed while singing.

In Hugo's novel, Gavroche's heroic actions are marvels. If, as Margaret Mead once opined, adults viewed children as "pygmies among giants,"[20] Hugo turns this vision on its head, describing the diminutive Gavroche as a giant concealed in a pygmy body and comparing him to Antaeus, the great mythical Libyan giant defeated by Hercules. As Hugo put it,

> The rebels watched with breathless anxiety. The barricade trembled, and he sang. He was neither child nor man but puckish sprite, a dwarf, it seemed, invulnerable in battle. The bullets pursued him but he was more agile than they. The urchin played his game of hide and seek with death, and . . . tweaked its nose.[21]

Gavroche does not survive, but when he is finally brought down by a bullet, Hugo tells us that "his gallant soul had fled."[22] For Hugo and for others, the child fighter very much represented "the people" in their struggle for democracy; in this sense, the child served as a collective representation of all that was good, striking to break out of an encrusted social order.

Hugo's story of the death of Gavroche must be placed in the context of his understanding of the violence of war. *Les Miserables* combines both narrative and social commentary and is marked by Hugo's observations on revolution, which he understood as inevitably flowing from the conditions of inequality in society. Hugo likened revolt to the releasing of a spring or, even more powerfully, to a whirlwind whose destructive force smashes those whom it carries away as well as those whom it seeks to destroy. It pulls in all those who cherish in their souls a secret grudge against some action of the state, life, or destiny, to the revolt; and when it manifests itself, they shiver and feel themselves uplifted by the tempest.[23]

Hugo carefully distinguished his judgments about the morality of collective violence from the particular makeup of the participants. The latter, he recognized, could be a rather motley crew of combatants. Hugo was well aware that violence could also take a negative turn. But, citing Lafayette, Hugo argued that true insurrection, as a form of expression of collective and universal sovereignty guided by truth, was a sacred duty.[24]

For Hugo, Gavroche's participation in the insurrection is part of the rights and duties of all citizens, men, women, and children, to resist oppression. Given the oppressive nature of childhood for children of his class background, Gavroche's best interests are served by participation in insurrection. In no sense could it be said of Gavroche that war "robbed him of his childhood," to use a modern humanitarian cliché. Instead, insurrection is the harbinger of a new moral order designed to eliminate the immorality of the social order that framed the ordinary life of a street child in nineteenth-century Paris.

Hugo sees a moral order in revolutionary violence. Indeed, because insurrection is a noble striving, revolutionaries must not act like criminals.[25] This view is made clear in an incident involving the murder of an elderly man, a doorkeeper who refuses a group of fighters entrance to a home. What seems clear is that Hugo is tracking the customary laws of war, which criminalize the intentional killing of noncombatants. Revolutions must follow the moral and normative codes of organized violence. But it is also clear that Hugo does not imagine that these would bar children from joining in class struggle.

The same revolutionary spirit that informs Hugo's novel is found in *Johnny Tremain*, one of the best-selling American novels for youngsters in U.S. history. Written in 1943, in the middle of World War II, it focuses on the saga of its eponymous hero as he grows and develops from a self-centered and arrogant child into a young soldier who takes up arms on behalf of the American Revolution. Johnny is 14 when the novel opens in 1773 and just 16 when it ends in the aftermath of the battles at Lexington and Concord, two of the most iconic events of the revolutionary era.

What is profoundly interesting is that the arc of Johnny's development and growing maturity tracks the arc of his emergence as a revolutionary activist. At the outset of the novel, Johnny is an apprentice to a silversmith in Boston and his main concerns focus on developing his abilities in a skilled trade. The novel adeptly recognizes that the American Revolution was very much a civil war that pitted loyalists to Britain (Tories) against the rebels (Whigs or patriots). Boston is a city divided between these groups, with many individuals holding feelings in between. Johnny himself is divided in his sentiments and in his conflicting love interest in both Priscilla, the patriot, and Lavinia, the Tory.

Johnny's transformation takes place after a prank by another apprentice results in the severe burning of his hand, which makes it impossible for him to continue on as a silversmith. Unable to find other skilled work, he is befriended by the 16-year-old Rab, whose family publishes and distributes a Whig newspaper. They hire Johnny to distribute the newspaper to its subscribers by horse throughout Boston and the surrounding areas. This brings Johnny into contact with various rebel leaders and groups, and this, together with his self-education in the library, turns him into a Patriot. As the rebel movement grows and the British military occupation deepens, Johnny is drawn into the violence of the growing rebellion. Several key events mark the transition to open revolt. These include a powerful patriotic speech by Otis, one of the Patriots; a cruel injury to Rab by a British officer who catches him trying to examine the locket of a musket; the execution of Pumpkin, a young British deserter who has given Rab his musket; and, finally, the death of the heroic Rab, who is fatally wounded at Lexington and who gives Johnny his musket just before he dies. Most importantly, it is discovered that Johnny's own burned hand can be made usable by a simple surgery and that, while he may never be a silversmith, he will be able to fire a musket in battle. In the last lines of the novel, Johnny also recalls Otis's speech and says, "Hundreds would die but not the thing they died for."[26]

The relationship between war, revolution, and the novel's construction of Johnny's development shows how political and revolutionary activity, including revolutionary violence, contributed to Johnny's development as a mature and responsible person. There is very little depiction of those things that are deleterious about war. Likewise, there is no mourning of the loss of Johnny's childhood. His life as a young teen is portrayed as constricted and confined by the narrow and dull system of apprenticeship. In this novel, it is revolution and the idea of fighting for an ideal that is seen as enhancing the individual and bringing the person from the narrow confines of childhood into the open vistas of adulthood.

The links between revolutionary violence, political maturity, social justice, and the transition to citizenship are no longer themes in contemporary novels of child soldiers although, to some degree, the heroic child soldier lives on in novels, usually set in a distant historical era, directed at adolescents. Contemporary contribution to this genre include Carol Campbell's *The Powder Monkey* and Arthur Trout's *Drumbeat*, all set in the American Civil War, but even some of these, such as *Soldier's Heart* by Gary Paulson, focus far more on the trauma of war than on its heroics.[27] But the heroic child soldier of the earlier era has been rendered invisible in contemporary adult fiction. Exemplifying this trend are three modern works of literature that reflect on the plight of child soldiers in Africa. All are published for the Western market. They are among the few works in fiction that give center stage to the actions of children under arms.[28] The works are *Beasts of No Nations* by Uzodinma Iweala, *Moses, Citizen and Me* by Delia Jarrett-Macauley, and *Johnny Mad Dog* by Emmanel Dongala.[29] In all of these works, the role of the child soldier is at best a terrible tragedy and at worst a threat to any sense of morality and social justice. Indeed the contemporary child soldier appears to subvert not just the social order but the natural order as well.

Beasts of No Nation is written as a comic nightmare allegory. As the title implies, it functions as the antithesis of a war novel. It is not about the human soldiers of a particular nation state but rather of "beasts" who have no national identification. Iweala was born in Nigeria but was educated in the United States and was named by Granta in 2007 as one of the twenty best American novelists. It is tempting to imagine that this story is set in Nigeria although the narrative does not follow any known conflict in Nigeria. Rather, it is a symbolic tale of modern warfare. The book tells a horrific story of the forced recruitment of Agu, a child soldier. It follows Agu through his initiation into the most brutal forms of violence: his participation in the gruesome murders of captured soldiers and civilians, which are portrayed in graphic detail, his drug-infused killing frenzies, and his routine rape and sodomization by the commander of his unit.

The book is set in a kind of dream time although, in this instance, the dream is a nightmare. From the very beginning, it makes use of the conventions of comic books. The Commandant is the nefarious nameless leader of the nameless force that murders Agu's father and kidnaps him from his village. The Commandant has all the attributes of a comic super-villain. Like the Joker in the Batman comics, he has no ideology. He is not interested in power, money, or land. He kills for the sake of killing as well as for his own lust and amusement. Like other super-villains, he has his servile minions, such as Luftenant and Rambo as well as his army of soldiers who laugh when he laughs and seek to imitate his every walk and gesture.

The terrible action scenes of the text are garnished with the classic devices of the comic-book narrative. In conventional comics, uppercase words such as ZAP, WHAM, BANG, and especially KAPOW mark the scenes of violence. In this book, Iweala converts and expands the classic KAPOW into a new faux-African action-comic vocabulary of evil: "KPAWA" marks the beating of Agu as he is dragged before the Commandant[30]; "KPWISHA," as the Commandant dashes cold water over him; "KPWUDA," as the machete wielding Agu chops a captured enemy soldier into pieces; "KPWUD," as he stomps a young girl to death; "KPWAMA" as soldiers kick down a door; "KEHI KEHI" marks the raucous laughter of soldiers as innocents are mutilated and murdered; and "AYEEEIII!" the murdered scream as they die. All this is very effective dramatically and none of it is funny in any way. But it has the immediate effect of stripping the story of any social and cultural context. The story unfolds both nowhere and everywhere. There is no history and no meaning to anything that is going on. Strongly paralleling the humanitarian understanding of war, it portrays people simply dying for nothing. Unlike the classic comic, there is no superhero to save the day.

Placing the action of the novel outside of any temporal, historical, or societal context gives the horror it describes an elusive transcendence. The action, which stands outside of history, stands for "Every War," or at least every African war, and in this respect at least, there is little difference between this novel's understanding of Africa and that found in Joseph Conrad's *Heart of Darkness*. Conrad took Africans out of history and suspended them between the human and the animal. Here, the narrator Marlow describes his journey up the Congo River:

> The pre-historic man was cursing us, praying to us, welcoming us—who could tell? We were cut off from the comprehension of our surroundings. . . . We could not understand because we were too far and could not remember because we were traveling in the night of first ages, of those ages that are gone, leaving hardly a sign—and no memories. . . . No, they were not inhuman. . . . They howled and leaped, and spun, and made horrid faces; but what thrilled you was just the thought of their humanity—like yours—the thought of your remote kinship with this wild and passionate uproar.[31]

And here is Agu, more than one hundred years later, on his way to the killing fields:

> We are walking down into the valley and down into the bush so I am feeling like an animal going back to his home. . . . I am hearing water and I am thirsty and wanting to drink. . . . Everybody is looking like one kind of

animal, no more human. . . . Everything is just looking like one kind of animal. . . . I am liking how the gun is shooting and the knife is chopping. I am liking to see people running from me and people screaming for me when I am killing them and taking their blood. I am liking to kill.[32]

In his influential critique of Conrad, Chinua Achebe decried Conrad's stripping of Africans of their humanity as well as his description of Africa as a "metaphysical battlefield devoid of all recognizable humanity."[33] Like Conrad's, Iweala's characters hover between the human and the inhuman, but the battlefield is no longer merely metaphysical. Still, the metaphysical struggle continues. Indeed, Iweala may well have a larger purpose in stripping his characters of their humanity, because in doing so he also immunizes them from their culpability in murder, a central theme in the humanitarian efforts to "protect" children under arms. But Conrad also had a larger purpose, namely, to offer a critique of colonialism. Yet, as Achebe tells us, "You cannot diminish a people's humanity and defend them" at one and the same time.[34]

Moses, Citizen and Me was written by Delia Jarrett-Macauley, who was born and resides in England but is of Sierra Leonean descent. The setting of the novel is the civil war in Sierra Leone. The novel's protagonist, Julia, living in London, is summoned back to Sierra Leone after a 20-year absence. She returns to the home of her beloved Uncle Moses and Auntie Adele in Freetown, where she encounters their grandchild, Citizen, an ex-child soldier who, during the civil war, murdered his own grandmother, Adele. Citizen, aged eight, is living with Uncle Moses after having being released from Doria, a rehabilitation camp for former child soldiers. Uncle Moses is torn between his grief for his murdered wife and his duty toward his grandson. The questions of the story are basic. Is Citizen ruined? Is he redeemable? Who can redeem him and how?

At first wanting to understand who Citizen is and later wanting to build a connection to him, Julia visits Camp Doria, where she has her first encounter with an ex-child soldier. The soldier, a boy nicknamed Corporal Kalashnikov, has just been rehabilitated from his regular habit of drinking tea laced with gunpowder and marijuana. Julia perceives him as caricature of a soldier who could otherwise be leading a carnival parade. Nonetheless, the encounter with Corporal Kalashnikov serves as a personal rite of passage, which enables her to begin to understand the plight of child soldiers.[35]

All of Julia's future contacts with child soldiers take place not in reality but in a magical dream-like state while her hair is being plaited by Anita, Uncle Moses's next-door neighbor. As Anita plaits her hair, Julia begins a magical journey into the forest where she encounters a unit of child soldiers that includes 12-year-old Abu, his older brother Masa, Citizen himself, and their vicious commander, the 20-year-old Lieutenant Ibrahim. Ibrahim is the leader of the "number-one-burn-house-unit."[36] The scene is one of stark brutality and violence. Ibrahim carries a knife that he has stolen from a corpse, one of many in the trail of corpses he has created in his campaign of extermination. The unit is about to attack a village of the "enemy," although it is clear that it is merely a rural village. The child soldiers, under the influence of drugs, join in the spree of chaos and murder, where the dying and fleeing inhabitants are seen by them as just so many insects. Despite their participation in murder, the children in the novel are presented as being completely under the murderous control of Ibrahim, whose calculated terrorism and violence propel them into combat. Ibrahim cruelly beats Abu for crying for his mother, lashes Citizen fifty times for his "failure" to beat a fellow child soldier to death, murders the helpless Musa who has come down with malaria, and forces the children to dance to stop them from comforting one another over Musa's murder. Despite this and despite how broken the children are, they retain their "innocence" and their humanity in this war.

For the rest of the novel, Julia magically tracks back and forth between Freetown and "the bush" where this unit of child soldiers, now deep in the Gola forest on the borderlands between Sierra Leone and Liberia, finds its redemption. The war is now over and the children are cared for by Bemba G, an elderly shaman-like character with magical powers. Bemba G's plan is to redeem and rehabilitate the child soldiers through the staging of Shakespeare's drama *Julius Caesar*, in which all the children will play a part. Citizen is to play Lucius, the boy servant of Brutus, who in the play is implicitly with Brutus when he dies on the plains of Phillipi and whom the novel casts as a boy soldier of ancient times. Lucius is sleeping in Brutus's tent when he encounters the ghost of Caesar, who foretells his death. In Shakespeare's play, Lucius cries out in his sleep, clearly disturbed by the presence of Caesar's ghost, but does not see him. In the novel, Citizen/Lucius, while acting out his part, has an unscripted dream where the ghost is not that of Caesar but of his murdered Aunt Adele. The encounter is transformative. Lucius sees the ghost and "the glory of her voice, those assessing eyes, naked brown arms with flesh gently drooping. He thinks of tenderness and love—and joining hands. The Ghost turns, revealing a back torn with wounds from a cruel death."[37] But in contrast to Brutus's encounter, Citizen's encounter with the ghost foreshadows not his doom but rather his reconciliation both with his family and with society.

To its credit, Jarrett-Macauley's novel does not seek to redeem child soldiers by members of the so-called helping professions—social workers and psychologists—but rather by reconnecting these child soldiers, who have been artificially isolated and brutalized by war, back into the global culture they have always inhabited. However, though the novel clearly demonstrates the intellectual richness of Sierra Leone society, its portraits of children at arms remain remarkably thin; Jarrett-Macauley reduces them to the stereotypes of human rights reporting. Indeed, as the author admits, she has never met a child soldier and has relied almost entirely upon her own interviews with personnel from agencies that deal with child soldiers. This is not to argue that a novelist must be constrained by reality but rather that the novelist's imagination, even in this otherwise wonderfully imagined story, has been constrained by the rhetoric of advocacy. This is a rather surprising result since Brutus's kindness and gentility toward Lucius on the very eve of his death suggest that Lucius and Citizen, despite their both being called child soldiers, actually have very little in common. Brutus may have betrayed Caesar but he is no Lieutenant Ibrahim. In the novel we never get a child soldier who departs from the stereotype. We never get the child soldiers who believed, even if wrongly, that they were fighting for a cause or those who fought to protect their homes and villages from rebel deprivations.[38]

Johnny Mad Dog by Emmanel Dongala is the story of a civil war in an unnamed country in Africa. It is partly based on the personal experiences of Dongala, who was a director of academic affairs at the University of Brazzaville. He fled the Congo at the onset of the civil war in 1997. *Johnny Mad Dog* is the story of two 16-year-old teenagers, Laokolo, a young girl on the run from the conflict who wheels her crippled mother around in a wheel barrow, and Johnny Mad Dog, a leader of a unit of a militia group called the Mata Mata, or Death Dealers. Johnny lives in a world of false-hood and deception. "Looting" he says, "was the main reason we were fighting. To line our pockets. To become adults. To have all the women we wanted. To wield the power of a gun. To be rulers of the world. . . . But our leaders and our president ordered us. . . . [to say] that we were fighting for freedom and democracy."[39]

Johnny Mad Dog's world is also one of total self-deception. He imagines that he brings sexual pleasure to a woman he is raping, regards himself as an intellectual even though he has only finished the second grade, and provides himself endless justifications for wanton murder. If this was a novel about a single individual, Johnny would clearly be a criminal sociopath. He is dangerous, glib, and grandiose; has absolutely no conception of the rights of others; and shows no guilt, shame, or remorse. But the essence of a sociopathic personality disorder is a disregard for cultural and social norms or rules. This novel portrays the world that Johnny lives in as itself devoid of meaningful social and political categories. The novel's use of patently absurd and inauthentic social and political categories conveys the meaningless cruelty of war. There are no authentic rules to break, which renders both Johnny and warfare ultimately unintelligible.

The novel uses a variety of rhetorical devices to do this. For example, the two ethnic groups at war are the Dogo-Mayi and the Mayi-Dogo, patently fictional ethnic categories that have blossomed out of squabbles between postcolonial political leaders with little prewar intergroup salience. The warring political parties formed around these ethnic categories are the equally contrived MFTLP (Movement for the Total Liberation of the People) and the MFDLP (Movement for the Democratic Liberation of the People). The categories of the opposition's allies are all a jumble, as Johnny and his militiamen imagine they are also hunting down fantasy "Chechens" and "Israelis," all of whom turn out to be innocent African civilians, who are casually murdered by Johnny and his unit. Some of this is balanced by the story of Laokolo, a courageous young woman of uncommon intelligence who tries to survive in an insane world. In the end, in an almost Orwellian way, both Johnny and Laokolo are manipulated by forces out of their control, and adults serve as stand-ins for Big Brother.

In the literature, folklore, and song about war, the very common name "Johnny" has frequently been used to mean every anonymous soldier. Over the last two or three centuries, there have been many Johnnys. Johnny Reb was the slang term for the common soldier of the Confederacy in the American Civil War. In the same war, soldiers of both the North and the South sang and marched to Patrick Gilmour's "When Johnny Comes Marching Home Again, Hurrah, Hurrah."[40] In World War I, Americans sang to "Over There," by George M. Cohan whose first verse began with "Johnny Get Your Gun, Get Your Gun, Get Your Gun," and in the World War II, U.S. audiences listened to the patriotic sounds of the Andrews Sisters singing Don Raye and Gene De Paul's "Johnny Get Your Gun, Again" in the 1942 film *Private Buckaroo*.[41] In all of these wars, especially the American Civil War, there were large numbers of children under arms, and "Johnny" easily stands for any soldier, whether adult or child. Certainly, Johnny Tremain, the eighteenth-century Johnny who was clearly a child soldier, fits easily into the genre of a patriotic soldier fighting for a just cause.

Of course, literature does not only provide us with patriotic Johnnies. Indeed one of the most powerful portraits of a "Johnny" in modern literature is the antiwar novel *Johnny Got His Gun* by Dalton Trumbo. The novel tells the tale of Joe Bonham, a young American soldier of the World War I who, after being hit by an artillery shell, lost his eyes and his limbs. Lying in hospital, Bonham is unable to communicate except by using his head to bang out Morse code.[42] At first blush, *Johnny Got His Gun* seems to speak of the unjustness of any war and to reject all attempts to justify war or to distinguish between just and unjust wars. The sheer horror of Bonham's situation implies that war is meaningless. Bonham has little use for any of the myths of war and rejects all of the so-called reasons for fighting—liberty, freedom, decency, democracy, or independence. Indeed, his thoughts on the American Revolution subvert every sentiment found in Johnny Tremain:

America fought a war for liberty in 1776. Lots of guys died. And in the end does America have any more liberty than Canada or Australia who didn't fight at all? . . . Can you look at a guy and say he's an American who fought for his liberty and anybody can see he's a very different guy from a Canadian who didn't? No by god you can't and that's that. So maybe a lot of guys with wives and kids died in 1776 when they didn't need to die at all.[43]

By the end of the novel we are less certain that Joe's position involves a complete rejection of war because the novel's critique of war is tied to a broader critique of class-based societies that locates the meaninglessness of war in an economic system that exploits the vulnerable. Although Bonham's acute suffering leads him to an antiwar position, it is by no means clear that Trumbo meant the novel to lead to a complete abandonment of the possibility of a just war. Indeed, Trumbo delayed the 1939 release of his book because he apparently feared it might unfavorably distort the efforts to defeat fascism in Europe. Thus, even the most powerful of antiwar novels demonstrates that all pro- and antiwar sentiments are coated in political residue.[44]

In stark contrast, recent novels of war and, especially, novels of children at war in contemporary conflicts completely remove war from the world of politics. None of these new texts of war offer a rationale for violence. Instead, war appears virtually out of nowhere, usually as a result of adult perfidy, to engulf children and to turn them into victims and killers. It is almost as if war was a malevolent natural phenomenon akin to a tornado, which lands on a country and destroys it. The novels attribute a kind of random and feral meaninglessness to war that unmistakably echo Conradian representations of the near-riotous inhumanity of Africans. It is not as if past wars and uprisings in the West, especially civil wars and revolutions, did not have dramatic displays of violence. Chateaubriand, in his memoirs, for example, describes terrible scenes of murder and mayhem during the French Revolution that are hardly supportive of Hugo's view of the morality of revolutionary violence. Chateaubriand described crowds of people bearing severed heads on spikes:

A troop of ragamuffins appeared at one end of the street . . . As they came nearer, we made out two disheveled and disfigured heads . . . each at the end of a pike. . . . The murderers stopped in front of me and stretched their pikes up towards me, singing, dancing and jumping up in order to bring the pale effigies closer to my face. One eye in one of these heads had started out of its socket and was hanging down on the dead man's face; the pike was projecting through the open mouth, the teeth of which were biting on the iron.[45]

Similarly, Chateaubriand's memoirs of the July Revolution, 1830, the same one in which Delacroix's painting figures so prominently, are unequivocal in their near-racialized disparagement of children and his horror at how they threw themselves into the bloody work of war:

> The children, fearless because they knew no better, played a sad role during those three days. Hiding behind their weakness, they fired at point-blank range at the officers who opposed them. Modern weapons put death in the hands of the feeblest. These ugly and sickly monkeys, cruel and perverse, immoral even without the capacity to perform immorally, these three-day heroes devoted themselves to murder with all the abandon of true innocents.[46]

Chateaubriand was a royalist and foe of revolutionary violence. His scorn for children under arms did not prevail in either French or Western thought, where the democratic gains brought about through revolution trumped virtually all other considerations. Thus, despite the cruel bloodletting of the past and the prominent role played by young people in revolutionary violence, revolutionary activity was understood as meaningful and positive. Yet, if Chateaubriand were alive today, he could easily be writing much of the contemporary humanitarian discourse on child soldiers.

The irony of why we were so willing to read a political and social context into the violent acts of children in the past but strip away this context in the present still remains. Why is it that we read mindless barbarism into contemporary warfare? Some might argue that the new wars in Africa and elsewhere are, in fact, much more horrible than the warfare of the past. The fact that war is increasingly directed toward civilians obviously adds to our sense of fear and outrage. To be sure, the portraits of African children at war that form the set-piece humanitarian and literary descriptions of child soldiers have been harnessed to serve modern notions of the greater good—ending children's involvement in war. But despite attempts to lend the situation of child soldiers a universal "everyman" quality, humanitarian and literary portraits of child soldiers do so by drawing upon an earlier discourse about Africa that served to dehumanize Africans. In the end, we are still writing Africa's script, and with it the larger story of child soldiers, in much the same way that Conrad did so many years ago.

Notes

1. Phillip Aries, *Centuries of Childhood: A Social History of Family Life*. New York, 1962, p. 411.
2. Frank Musgrove, *Youth and the Social Order*. London, 1964.
3. Aries, *Centuries of Childhood*, pp. 202–206.
4. Ibid., pp. 266–268.
5. Peter Gripton, *The Arborfield Apprentice*. Reading, 2003.
6. Herman Melville, "The March into Virginia Ending in Manassas," in Lorrie Goldensohn, ed., *American War Poetry: An Anthology*. New York, 2006, pp. 65–66.
7. Richard Reid, *War in Pre-Colonial Eastern Africa*. London, 2007, pp. 2–21.
8. Anna Freud and Dorothy Burlingame, *War and Children*. New York, 1943.
9. Most "child soldiers" are adolescents, who are legally defined as children.
10. "Child Soldiers Ratification Campaign," in *Human Rights Watch: Children's Rights*. http://www.humanrightswatch.org/campaigns/crp/action/index.htm. Accessed 23 November 2007.
11. Sally Merry, "Human Rights Law and the Demonization of Culture (And Anthropology Along the Way)," *Polar: Political and Legal Anthropology Review*. 26, 1, 1999, pp. 55–77.
12. For example, Francis Deng, *The Dinka of the Sudan*. Prospect Heights, IL, 1972, pp. 68–73; E. Adamson Hoebel, *The Cheyennes*. New York, 1978, p. 77.
13. A. W. Cockerill, *Sons of the Brave*. London, 1984.
14. John C. Dann, *The Revolution Remembered*. Chicago, 1980.
15. David B. Parker and Alan Freeman, "David Bailey Freeman," *Cartersville Magazine*. Spring, 2001. http://www.wintektx.com/freeman/whois.htm. Accessed 28 October 2007.
16. Margaret Downie Banks, "Avery Brown (1852–1904), Musician: America's Youngest Civil War Soldier," *America's Shrine to Music Newsletter*. February 2001. http://www.usd.edu:80/smm/AveryBrown.html. Accessed 28 October 2007.
17. Susan Hull, *Boy Soldiers of the Confederacy*. Austin, TX, 1998 (1905).
18. Delacroix's painting is available on: Wikipedia. http://fr.wikipedia.org/wiki/Eug%C3%A8ne_Delacroix. Accessed 28 October 2007.
19. Jean-Jacques Yvorel, "De Delacroix à Poulbot, l'image du gamin de Paris," *RHEI: Revue d'histoire de l'enfance irrégulière*. 4, 2002. http://rhei.revues.org/document52.html. Accessed 23 May 2007.
20. Margaret Mead and Martha Wolfenstein, *Childhood in Contemporary Cultures*. Chicago, 1955, p. 7.
21. Victor Hugo, *Les Miserables*. New York, 1976 (1862), p. 1028.
22. Ibid., p. 1028.
23. Ibid., p. 883.
24. Ibid., p. 887.
25. Ibid., p. 939.
26. Edna Forbes, *Johnny Tremain*. New York, 1980, p. 256.

27. Carol Campbell, *The Powder Monkey*. Shippensburg, PA, 1999; Gary Paulsen, *Soldier's Heart*. New York, 1998; Arthur Trout, *Drumbeat*. Shippensburg, PA, 2007.

28. Other works of fiction contain episodes in which child soldiers appear, including Chimamanda Ngozi Adichi, *Half of a Yellow Sun*. New York, 2006; Helon Habila, *Measuring Time*. New York, 2007; Ahmadou Kourouma, *Allah Is Not Obliged*. New York, 2006.

29. Uzodinma Iweala, *Beasts of No Nation*. New York, 2007; Delia Jarrett-Macauley, *Moses, Citizen and Me*. London, 2005; Emmanuel Dongala, *Johnny Mad Dog*. New York, 2006.

30. Iweala, *Beasts of No Nation,* pp. 2, 8, 21, 47, 51.

31. Joseph Conrad, *The Heart of Darkness*. Clayton, DE, 2004, p. 37.

32. Iweala, *Beasts of No Nation,* p. 45.

33. Chinua Achebe, "An Image of Africa: Racism in Conrad's *Heart of Darkness,*" in Gregory Castle, ed., *Postcolonial Discourse: An Anthology*. London, 2001, pp. 209–220.

34. Caryl Phillips, "Out of Africa," *The Guardian*. 22 February 2003. http://books.guardian.co.uk/review/story/0,12084,900102,00.html#article_continue. Accessed 21 November 2007.

35. Jarrett-Macauley, *Moses, Citizen and Me,* pp. 36–37.

36. Ibid., pp. 58–61.

37. Ibid., p. 208.

38. David Rosen, *Armies of the Young: Child Soldiers in War and Terrorism*. New Brunswick, NJ, 2005; Paul Richards, *Fighting for the Rain Forest*. Portsmouth, NH, 1996.

39. Dongala, *Johnny Mad Dog,* p. 64.

40. Patrick S. Gilmore, "When Johnny Comes Marching Home Again Hurrah, Hurrah" (1863), in *Patriotic Melodies*. United States Library of Congress. http://lcweb2.loc.gov/diglib/ihas/loc.natlib.ihas.200000024/default.html. Accessed 6 December 2007.

41. George M. Cohan, "Over There" (1917), in *Patriotic Music*. United States Library of Congress. http://lcweb2.loc.gov/diglib/ihas/loc.natlib.ihas.200000015/default.html. Accessed 6 December 2007; Edward F. Cline, dir., *Private Buckaroo*, 1942; Internet Movie Data Base http://www.imdb.com/title/tt0035218/soundtrack. Accessed 6 December 2007.

42. Dalton Trumbo, *Johnny Got His Gun* (New York, 2007).

43. Ibid., p. 145.

44. Ibid., p. 2

45. Robert Baldick, *The Memoires of Chateaubriand* (New York, 1961), p. 105.

46. François René de Chateaubriand, *Mémoires d'outre-tomb*. Paris, 1951 (1841), p. 430, cited in Yvorel, "De Delacroix à Poulbot, l'image du gamin de Paris." I thank my friend and colleague Richard Rabinowitz, of the American History Workshop, for his translation of Chateaubriand.

CHAPTER 6

From Bedpans to Bulldogs: *Lottie: Gallipoli Nurse* and the Pitfalls of Presenting War to the Young*

Sara Buttsworth

Upon visiting the Auckland War Memorial Museum for the first time, I was struck by what I found in the bookshop. Gazing out at me, framed by a starched white veil, was *Lottie: Gallipoli Nurse,*[1] and not far from her on the shelf was the jowled grin of *Caesar the Anzac Dog,*[2] looking hopefully down at a wounded soldier. Both these books have direct connections to the museum. Charlotte Le Gallais' story (*Lottie*) and photograph along with photographs and diary extracts pertaining to her brothers Owen and Leddra have prominence in the main museum exhibit on the Great War, while Caesar's collar can be viewed in different places depending on display space. The impact of these books, however, is not only due to their connections to the "real" in the museum but also as part of a growing body of picture books aimed at New Zealand children from ages seven through 14 that re-present and reconstruct First World War mythologies. Initially, I experienced pleasant surprise at the visible visual representation of women in wartime that is superficially presented by *Lottie: Gallipoli Nurse.* In retrospect, what should not have astonished me was the ways in which this publication, connected to a place of popular memory and memorial, upheld the combat-centric ideology of ANZAC (an acronym for the Australian and New Zealand Army Corps).

The First World War still dominates the memorial landscape of New Zealand, and the battles in which New Zealand soldiers took part are considered by many to be the nation's "baptism by fire." ANZAC fought under British Command in the First World War and is largely associated with the failed Gallipoli campaign of 1915. "ANZAC," the acronym, was used to specify the army corps but is now used to symbolize military heroism in Australia and New Zealand from that conflict right up to the present. It also has common usage with reference to sporting events between teams from the two countries, particularly rugby, rugby league, and cricket. Whatever the endeavor, the heroism of ANZAC is always masculine and combatant, and the focus of the legend is that heroism rather than the historical intricacies of the context or legacies of the Great War. The "tall bronzed ANZAC" dominates the way in which the First World War is remembered and has until recently left very little room for examinations of New Zealand society and the frictions that continued between different sectors because, or in spite, of the war. There has been little room for the discussion of noncombatant participants in the war, particularly women, in the public discourses of memory and memorialization.

ANZAC Day memorial services and ceremonies are held on 25 April every year to commemorate the bravery of New Zealand soldiers and the severe casualties they suffered during the many futile battles of the First World War but especially at Gallipoli. The government and many of New Zealand's people greeted the First World War as an opportunity for the fledgling nation to prove itself on the international stage. In all, 18,166 New Zealand soldiers lost their lives, and there were over 55,000 casualties throughout the duration of the conflict.[3] The "great adventure" became both a cause for national mourning and a focus of national pride. The treatment of the defeat at Gallipoli by New Zealand media and later historians formed a pattern where devastating losses were deemed a testament to the bravery and skill of New Zealand soldiers, who suffered at the hands of an incompetent leadership and insurmountable odds. The battle for Passchendaele in France in 1917 saw larger losses of life of New Zealand soldiers than the debacle at Gallipoli two years earlier, but it is recounted in similar ways: a place where New Zealanders proved their worth against impossible odds and a deficient (British) general staff. These battles are by no means the only places New Zealand soldiers fought in the First World War, but they are the most prominently remembered in the late twentieth and early twenty-first centuries.

New Zealand's war culture focuses almost exclusively on the combat soldier, a telescoped vision that has come at the expense of broader local, national, and international historical contexts. It represents vignettes of the battlefield almost as if they were the whole and only tenable picture of war. The focus on combatants in the field is exemplified in recent public debates over which battles were more important to New Zealand—as a nation— with body counts a key indicator of significance. The ninetieth anniversary of the Battle of Passchendaele in 2007, for example, put this focus on casualties as an indication of valor firmly in the spotlight when historian Glyn Harper, echoed by Prime Minister Helen Clark, called for this battle to be recognized in the same way as Gallipoli was, because more New Zealanders died at Passchendaele.[4] This kind of attention does not broaden the focus of historical or cultural inquiry away from the traditional images of war as a male pursuit (or something similar); it merely relocates it from the sand and flies of Turkey to the mud and trenches of France. New Zealand participants in this conflict who did not, or were not allowed to, take up arms and the small but vocal minorities who for diverse reasons opposed the war remain indistinct in the public image of the war. On ANZAC Day "lest we forget" only really applies to combatant soldiers.

Mythologies of war are often presented as if they are self-explanatory and as sites that resist rather than invite questioning. Historian Deborah Montgomerie has articulated many of the problems in dealing with New Zealand histories of the First World War as hinging on the replication of mythology by written histories. In reconnoitering the state of New Zealand's histories and historiography of war, Montgomerie points out that "war, we are told, brought us to self-consciousness as a nation, but the details of the process . . . remain indistinct."[5] The centrality of the carnage and defeat at Gallipoli in 1915 dominates that national "self-consciousness" and continues to function as "*memento mori*, heroic folk tale and political parable."[6] Similarly, the children's picture books under scrutiny in this chapter form a part of a broader ANZAC tradition that in Australia and New Zealand is frequently treated as sacred and hermetic. However, just as this is not true of the mythologies from which they draw their inspiration, these texts to a large extent cannot and do not operate holistically unto themselves. They are really only comprehensible within the broader framework of ANZAC mythology and broader stories about the First World War.

From within her starched white veil, *Lottie: Gallipoli Nurse* looks out to an audience of young people who do not have a living connection to the war that was to end all wars. Almost a century after the First World War began, a small but significant number of New Zealand picture books centering on that conflict have been published. The production of such texts raises questions both about and beyond their immediate subject matter and invites an evaluation of the presentation of a martial mythology to a peacetime, noncombatant, and young audience. How can, or should, war be presented for and to children? And to what end? What is the relationship of these texts to broader discourses on war and childhood? *Lottie* invites all of these questions and raises particular issues relating to narratives of war and the representation of gender. As a story that focuses on the experiences of a female nurse within a tradition that has little room for women's stories in both broader war narratives and the narrower scope of ANZAC mythology, *Lottie* has the potential to play an important role. A book about a nurse and a noncombatant is unusual in the New Zealand canon of historical literature about the First World War. I was initially intrigued and delighted to find what appeared to be such an unusual text, which I hoped might attempt to deal with a story too long overshadowed by mythologies of the heroic and doomed soldier. However, in spite of the fascinating story of Charlotte Le Gallais upon which the book is based and its promising title, *Lottie: Gallipoli Nurse* continues the sanctification of the combatant soldier and in subtle ways makes its own main character an understudy in her own performance. *Lottie* represents a lost opportunity, reinforcing rather than challenging the socially acceptable ideas and tropes about the war and the role of New Zealanders in it.

While Lottie's noncombatant status is unusual in New Zealand historiography, within the parameters of books aimed at children, however, she lines up with an array of different personalities that includes a number of animals—the dog in *Caesar the Anzac Dog*, the bantam of *The Bantam and the Soldier*, the donkey of *The Donkey Man*—but no other women. What a close examination of this text reveals is that in spite of first appearances, in both the broader context of historical literature and ANZAC mythology and the narrower scope of children's picture books, Lottie's presence is far from unproblematic in terms of narrative, visual, and historical representation. The problems posed by *Lottie: Gallipoli Nurse* require an interdisciplinary, intertextual, and intratextual approach. Charlotte Le Gallais was an independent professional woman who sailed vast distances to "do her bit" during the Great War. While obviously no text is an island, the contrasts between the life of Le Gallais and how it is represented in *Lottie* as well as the book's reliance on some prior knowledge and acceptance of the mythologies of the First World War are striking.

In New Zealand since the late 1990s, a number of picture books dealing with the First World War have been published for New Zealand young people. *The Bantam and the Soldier,* one of the best known, seems to have begun the trend in 1996.[7] In 1997, it won the prestigious New Zealand Post Book Awards' Children's Book of the Year. That same year, John Lockyer's *Harry and the Anzac Poppy*[8] was published, followed in 1998 by *Lottie: Gallipoli Nurse*. In conjunction with Chris Pugsley, a well-known historian of New Zealand and war, Lockyer was then involved in the production of *The Anzacs at Gallipoli*, a factual (rather than fictional or pseudofictional) account of the Gallipoli campaign, which was published in 1999.[9] *Caesar the Anzac Dog* by Patricia Stroud and Glyn Harper's

The Donkey Man followed in 2003 and 2004, respectively.[10] Furthermore, Jennifer Beck, the author of *The Bantam and the Soldier*, has most recently collaborated with Lindy Fisher in the writing and illustrating of *A Present from the Past*, which indirectly looks at the role of the Red Cross and nursing during the First World War through the interesting trope of gift boxes acting as receptacles of memory and memorabilia.[11] There have been repeated calls for children and young adults to be taught more about the First World War in New Zealand, with an emphasis on New Zealand's military history and its connection to "nation-building."[12] This focus on the First World War in the public culture of New Zealand was a part of the impetus for the publication of all these picture-book texts, including *Lottie: Gallipoli Nurse.*

Lottie is based on the experiences of Charlotte Le Gallais, who was one of 14 New Zealand nurses who sailed to Gallipoli on board the hospital ship *Maheno* in July 1915[13] and one of 550 nurses who served with the New Zealand Expeditionary Forces throughout the duration of the First World War.[14] Le Gallais was practicing as a registered nurse at Auckland hospital when the war broke out. In her early thirties and engaged to be married, Le Gallais enlisted in the New Zealand Army Nursing Service (NZANS) in 1915.[15] She served on the *Maheno* as it picked up wounded soldiers from the beaches of Gallipoli and ferried them to a hospital on the island of Lemnos. She visited Malta and spent some time in England as well. On her return to New Zealand, she married her fiancé Charles Gardner (who had spent the war years in New Zealand), and subsequently had two children. She died in 1956. Charlotte Le Gallais had four brothers, two of whom served in the New Zealand Expeditionary Forces during the war. Leddra fought in the Gallipoli campaign and was killed on 23 July 1915. Owen fought in France and returned home in very poor health as a result of his years of active service. *Lottie: Gallipoli Nurse* centralizes the relationship between Charlotte and Leddra and emphasizes Charlotte's concern for her brother's well-being and desperation for news of him. Perhaps in the interests of simplicity, Lockyer makes no reference to Owen, the other brothers, or the fiancé to whom Charlotte confided the details of her wartime service in letters.

Lottie: Gallipoli Nurse is, according to the author, John Lockyer, a "story based on letters Lottie and Leddie sent home and the war diary of John Duder, an officer based on the *Maheno*."[16] These materials are mentioned by Lockyer in a kind of afterword, in which he briefly provides information of the fatal Gallipoli campaign, including numbers of casualties, and a brief description of Charlotte Le Gallais who "was a nurse on the *Maheno* . . . eager to go to Gallipoli to care for the men but also to meet up with her brother Ledra [*sic*] (who had left New Zealand on a troop ship a few weeks before her). Sadly they never met. Ledra [*sic*] was killed at Gallipoli on 23 July 1915."[17] The accessibility and completeness of the archival material upon which *Lottie* is based was quite possibly a reason Lockyer chose Charlotte Le Gallais as the subject for his book.[18] In which case, it seems rather a pity that more was not done with the excellent materials Lockyer had available to him to turn what he labeled a work of "faction" into something more true to Charlotte's war experiences.[19] Much like the stories of old soldiers, the details of who Lottie was and what her life was like outside her military service are omitted so that war takes center stage. This tends to make her appear rather one dimensional, especially when considered in conjunction with the depiction of her war service purely as a result of her feelings of duty and love for her brother, the soldier.

The importance of Lockyer's *Lottie* lies in its use as a public text to represent and explore the lives of New Zealanders in the war. Significantly, Le Gallais' story opens up the possibility of investigating experiences that have largely been marginalized by traditional histories of New Zealand in the First World War—those of the noncombatant, but particularly female, population. Both a museum exhibit utilizing the Le Gallais family archive and *Lottie* attempt to go some way toward rectifying the absence of the small but important NZANS in the narration and memorializing of the First World War. The connection between Charlotte as a nurse and Lottie as a character is so important to the Auckland War Memorial Museum that on ANZAC Day in 2007 and 2008 an actor dressed as Lottie told Lottie's story in the vicinity of the exhibits that display the story of the combatant and noncombatant Le Gallais'. The sale of *Lottie* in the museum bookshop forged a further tangible connection between this "work of faction" and the materials upon which it is based. A blurring between the historical person and the fictional character has resulted. She has become, quite literally, a character narrating war directly to a young audience. Whether this means that the experiences of Charlotte Le Gallais and other women like her are read as fictional and, therefore, are undermined or that the audience views Lottie as a "real" person, lending historical weight to a book that has been deliberately fictionalized, is a confusion worth pondering.

This conflation of "real" and "fictional" has occurred partly as a consequence of the relationship between this picture book and the opportunities it provides for opening up the subject of the First World War for young people. Attempts to make this story relevant and comprehensible to a contemporary young audience have often been at the expense of being true to the source material and historical accuracy. There has been an assumption that young readers will have a vague familiarity with the myths of the

First World War and the importance of Gallipoli to New Zealand. Lockyer's *Lottie* can only be understood in relation to the broader contexts of war, gender, and ANZAC mythologies within which it was created and is read. Without careful explanation and interpretation, this book cannot illuminate the experience of war for its peacetime readers, relying as it does on the preexisting dominant narratives of the ANZAC legend. What it does not invite is a complex appreciation of stories that must be historicized, moving as they do from their original context and source material a century ago to their re-presentation in a much different time and place and to a much different audience. The contexts of Charlotte Le Gallais' story and its adaptations, therefore, shift uncomfortably between archive, exhibit, picture book, and classroom and are further blurred by the interplay—and lack of it—between the written word and pictorial content (discussed below). As Antoine De Baecque would have said, *Lottie* is not only intertextual but also "intercontextual."[20]

The use of the term "faction" implies a "based on a true story" approach with some details changed or tailored to suit a younger audience and the contemporary expectations of an educational text. Lockyer's written text utilizes archival material in many places word for word, which, in combination with the exhibits of photographs and letters in the war memorial museum, lends this picture book its authenticity and authority. Other picture-book texts, *The Bantam and the Soldier,* for example, employ photographs or depictions of memorabilia from the period as a means of "providing depth" or anchoring a narrative that is being constructed long after firsthand memory has receded.[21] *Lottie* does the exact opposite. Its illustrations, as we will see, lend most of the fictional aspects of this work of "faction" while anchoring its text in the "known" and "available." This is both a strength and a failing.

The closeness of the text to the archival material in places masks some of the omissions made by Lockyer to make this work "suitable" for children in the late twentieth century. The ugly side of war in general is not absent from *Lottie.* Debates on the exposure of children to violence through the media are too numerous to be discussed here, and the exposure of children to "gratuitous" violence continues to be a hot topic in many circles. The depiction of "real" or historical violence is, however, a necessary part of making sure that children are included in the endeavor to ensure war is not repeated. Exposure to depictions of real and imaginary violence, death, and destruction is inevitable for most children, even in a peaceful nation like New Zealand. In Western culture, violence and war go hand in hand. Lockyer takes pains to ensure that the difficulty of dealing with the dreadful wounds, disease, and piles of dead bodies is a part of Lottie's story.[22] He also replicates the uncertainty of a soldier's life, repeating Leddra's written comment to his father in a letter prior to his death:

> I never thought I would be a soldier but now I am one I am determined to be a good one, to do my duty to the best of my ability. If I have bad luck, well, I suppose it has to be.[23]

Interestingly, there is no manifestation here of the part of the ANZAC myth that assumes that Australian and New Zealand men were *natural* soldiers.[24] Rather, Leddra (and Lockyer by repeating him) hints at the doubts the former school teacher had, and this is an attempt to reassure both himself and his family that he can be a soldier and do his best for king and country.

While references to violence and death are not omitted in *Lottie*, the ugliness of early-twentieth-century racism and imperialism is missing. The Le Gallais correspondence is peppered with disparaging references to the "noisy" "natives" in Kandy and the "Arabs" in Egypt with all the arrogance and misunderstanding of a white middle-class female citizen of a far-flung corner of the British Empire. The contexts of empire and its inherent racism, which are such a central part of understanding New Zealand's participation in the First World War, are absent from the picture book that, instead, fits neatly into a contemporary discourse about the waste and destruction of the First World War and the reluctant heroism of those who fought in it. In this way, *Lottie* dovetails nicely with the contemporary international canon of children's literature about the First World War in which these themes are also largely missing.[25] The repulsiveness of death in the trenches represented in *Lottie* reinforces the futility of war. However, it does not help to explain why the war was fought nor does it illuminate the ongoing significance of the First World War to an audience whose knowledge of it is limited.

The "factional" elements in *Lottie* are further complicated by the book's proposed educational aim, partly as a result of the truth effect it carries through its status as a book "based on fact" and that is educational in intent. *Lottie* deviates from its original archival base in a number of ways. Lockyer may well have made some of these changes to elicit sympathy or promote understanding from his young late-twentieth-century audience, but the deviations potentially place the educational impact in jeopardy. This is perhaps best illustrated by what may seem as a minor alteration in the characterization of the *Maheno's* mascot, a bulldog named Jock, as a Scottish terrier. Perhaps, a Scottish terrier can be perceived as friendlier than a bulldog in the eyes of children already immersed in a culture that prizes visual appeal. Or maybe the author considered that the name Jock could only belong to a Scottish terrier, and the illustrator followed suit. Nevertheless, the alteration is fundamental, particularly as Jock plays a much larger role in *Lottie* than he does in the archival material. He is centralized in the book's narrative as Charlotte's source of comfort when she learns about her brother's death. The prominence of an animal is not uncommon in literature for children about war. Animals are often used as a means to soften the content and make it more accessible. Many animals are depicted in the other books mentioned above, and they also feature in children's literature throughout Western culture.[26] The centrality of an animal like Jock in *Lottie* helps to illustrate an ethic of caring and builds a bridge to experiences of grief young people may have had with regard to either a beloved pet or a family member.

However, the transformation of Jock from bulldog to cuddly terrier also serves to represent gender in a traditional way. It leads to the depiction of a young girl confiding in a small faithful pet rather than the portrayal of a mature woman who has no direct relationship with the ship's mascot at all and instead has close friendships with other women and a keen interest in medicine and professional care. The trans-dog-rification of Jock to elicit sympathy comes at the expense of historical accuracy and the possibility of telling a different kind of story about women and war (see figure 6.1).

More significant than the change in breed of Jock is that Lottie bonds with a dog in the book but is separated from her nursing comrades and the soldiers and sailors on board the *Maheno*. The depiction of her relationship with Jock is in keeping with ideas about young women, nursing, and an ethic of caring that are not far removed from late-nineteenth- and early-twentieth-century ideas about why nursing was an appropriate profession for women. Nursing was an extension of ideas and ideals about women's role in the home and broader society. However, perhaps more importantly, the portrayal of her relationship with a dog sets her apart from the other women around her rather than forging connections between them. She is isolated, a not unusual tactic in the representation of women in unconventional situations.[27] This lack of identification with a community of other women and a history of professionalism is particularly important when considering that *Lottie* has appeared and been reprinted in an era considered by some to be "postfeminist." In the postfeminist era, only individual action matters, and the legacies of feminism are frequently belittled or trivialized, which is particularly true in the popular culture in which the intended young audience of this text is immersed.[28] In this context, the relationship of a young woman and a dog becomes much more important than Lottie's experiences as a nurse or her interaction with other nurses.

Figure 6.1 Official postcard of the hospital ship *Maheno*. Depicted are New Zealand nurses on deck in uniform and Jock the bulldog, the ship's mascot. In *Lottie: Gallipoli Nurse*, Jock is depicted as a Scottish terrier.

Source: The New Zealand Hospital Ship "Maheno": First Voyage July 1915, to January 1916. Christchurch, 1916, p. 43.

Interestingly, Jock only receives passing mention in the actual letters of Charlotte Le Gallais. The source in which he is far more prominent is the official diary kept by John Duder, first officer aboard the *Maheno* and the other main source for Lockyer's book.[29] So, in the transposition of archival material to pictorial "faction," the "fine big bulldog," which was a close companion of a male ship's officer, became a small Scottie dog, a source of comfort to a pining girl. This change of breed and relationship with the dog is significant when considering the representation of gender to a modern audience and the conflict of the necessity of the portrayal of war with the imperative of dulling the harshness of its realities.

The complexities of any text are deepened when they target a young audience, accompanied as they are by the social expectations that imbue childhood. Texts for children about sacrosanct subjects such as war, death, or national heritage can receive close scrutiny from a (adult) society that fears that the text may interfere in the formation of that child's worldview or, worse, damage his or her "innocence." And once a text is accepted as child appropriate, regardless of its subject matter, it can attain its own sacrosanct status. Criticism or challenging of such texts is paramount to undermining the memories of childhood that the child is cultivating (or the nostalgia and memories of childhood that adults may have). In many respects, therefore, "acceptable," and even more so, "notable" texts become untouchable and their representations of the past (even if they are fictional or factional) become a sacred "reality." Such texts are as much part of the grand narratives on which they are founded as they are cultivators of the myths that underpin them.[30] So too the national mythologies of war— details about casualties and battle tactics—are often subject to close scrutiny, while the grand narratives that contain them can be considered so sacred as to be indestructible. In the late twentieth century, this book about a young woman yearning for her brother does far less to disrupt or even question the ANZAC tradition than a picture book about conscription, the treatment of conscientious objectors, deserters, prisoners of war, or the racism faced by the Maori battalion might have. While *Lottie* is unique in its subject matter, it reinforces rather than questions the preexisting ideas its audience (both children and adults) has.

Despite the promise of Lottie's steadfast gaze on the cover and the centrality of her position as "Gallipoli Nurse" in the title, this picture book concentrates much more on the absent brother, Leddie, than on the experiences of nurses during the First World War. In making Lottie the narrator of her brother's story rather than of her own history and centralizing the search for Leddie instead of the experiences of a group of professional women working under extraordinary circumstances, *Lottie: Gallipoli Nurse* functions as a part of the canon of ANZAC mythology. It does not tell a new story or invite a refashioning of the legend. In contrast, *The Bantam and the Soldier* subtly shifts the legend of ANZAC to tell the story of the loneliness of being a soldier in France and infers that the much-touted ideas of "mateship" were not universal. It also portrays combatants in a gentle light rather than one of grand heroism. *The Bantam and the Soldier* is also significant in that it attends to the continuities in the soldier's life and his return to the farm rather than living on in memory as one of the fallen. The focus on France is also important as it moves away from the centrality of the Gallipoli campaign. While *The Bantam and the Soldier* moves away from more traditional narrations of the First World War, it has also been the recipient of prestigious awards, perhaps indicating that, certainly in literary circles, it is important that different kinds of stories are considered. Unlike *Lottie* though, *The Bantam and the Soldier* does not carry with it the weight of archival connections or present the conflation of fact and fiction. *The Bantam and the Soldier* really stands alone in its movement away from traditional depictions of ANZAC mythology.

Lockyer's earlier work *Harry and the Anzac Poppy* also centers on France and has a peripheral view to the suffering of those left at home, but this view is not extended, and the conventional stories associated with New Zealand and the First World War continue.[31] *Harry*, set as it is in 1917 with a married soldier on the Western front in Europe its main protagonist, could have opened up the possibility of discussing complex issues like conscription and particularly the conscription of married men in New Zealand in 1917. Instead, this is merely a device through which to introduce a child. The story is structured around a young boy reading old letters his grandmother received from her father during the war. The introduction of not one but two children's perspectives was perhaps a means of bridging the century's divide between characters and intended audience. It certainly allowed the explanation of certain things like what "shells" were—no, not "egg shells," but bombs.[32] The focus on Harry as the main character rather than on his grandmother reinforces the generational aspects of ANZAC mythology that it is something that can be passed from father to *son*—the heirs of a masculine combatant tradition.

Where *Lottie* differs from *The Bantam and the Soldier* is in its lack of an active voice. *Lottie* reinforces traditional ideas about women and war—as the ones waiting for things to happen to their men, even if that waiting is not separated from the war zone by geography. This feminine narrative is not one of action or independence but one of sorrow and inactivity. The challenges of nursing in the early twentieth century are sidelined for a more conventional story in which the absence and death of a beloved brother take the spotlight. The complex motivations of a young professional woman receive little attention, and instead, the implication of *Lottie* is that the eponymous protagonist enlists to be close to her brother.[33] This may be a part of a device to gain the sympathy of young readers. It is not uncommon for close sibling relationships to be portrayed in children's literature, particularly when children or teenagers find themselves without adult protection.[34] However, Lockyer overuses this trope considerably. There is only one page in the entire book where a direct mention of hope or fear or sorrow regarding Leddie is absent, and even here, Lottie's enquiries about Gallipoli infer the search for her brother.[35] Leddra is present even in the discussion of nurses dying on a torpedoed ship or of the wonders of Egypt. The combatant who becomes a casualty is omnipresent in a story that is ostensibly about his noncombatant sister. This sidelining of the book's supposed main character is even more obvious in the afterword, which provides the reader with some cursory background to the Gallipoli campaign but not of the NZANS. Instead of providing some basic statistics on how many women served, Lockyer repeats the casualty statistics pertaining to the soldiers who lost their lives. The nurses who lost their lives are not mentioned, nor is the struggle New Zealand nurses faced to be allowed to serve their country in the first place. Despite the appearance of a different kind of narrative promised by the book's cover picture and title, they act as little more than a trompe l'oeil for a book reiterating martial masculine endeavor and the primacy of the roar of battle.

In a culture that values the minutiae of military history, another important change in the written text is the author's conflation of the well-known (in New Zealand) sinking of the *Marquette*, where ten nurses died, with the fate of another ship that had nothing to do with the NZANS, *The Royal George*. This confusion may have arisen as a result of Duder mentioning the torpedoing of a ship named *The Royal George* in his diary entry of 4 August 1915.[36] However, the sinking of the *Marquette* is a reasonably familiar incident to many New Zealanders, partly as a result of the deaths of some of the nurses on board. The *Marquette* was actually sunk on 23 October, and Charlotte Le Gallais mentions the incident in her letter of 17 November 1915.[37] A simple recognition of the dates would have avoided this conflation and confusion. It may, of course, have been a part of the author's "factionalisation." However, it highlights some of the real problems in using "faction" to educate children about history. In teaching the First World War to secondary school students aged between 12 and 15 in New Zealand, *Lottie* is frequently pointed to as a suitable learning resource. The New Zealand social studies curriculum available online through Unitec, for example, has a module called "Gallipoli Webquest," and *Lottie* is one of the recommended print resources as are the entirely fictional *Harry and the Anzac Poppy* and Ken Catran's *Letters from the Coffin Trenches*.[38] Nowhere on the website is there a guide to dealing with the differences between archival material, works written as scholarly history, and works intended as either "faction" or fiction. *Lottie*'s status as an educational resource surely requires it to represent factual accuracy as much as any other historical work, particularly if student and teacher resources do not problematize the different kinds of materials in use.[39]

The confusion this book presents has been further complicated by a historian of nursing using *Lottie: Gallipoli Nurse* as a reliable historical source. Anna Rogers, who has written a very important and, in all other respects, well-researched book on New Zealand army nursing, has been a victim of the "truth effect" of this work of "faction." Rather than sourcing archival material, she replicates the "factional" by citing Lockyer's incorrect description of the sinking of the *Royal George* in addition to her discussion of the actual sinking of the *Marquette*.[40] In a culture of war stories where the tiniest details are endlessly discussed and examined by military historians,[41] this kind of intertextual blurring between archive and picture book cannot but hinder the attempt to reflect accurately on the diverse experiences of women and war, let alone on Charlotte Le Gallais' real life.[42]

Where the blurring between fact and fiction is most problematic in *Lottie*, however, is in its pictorial content. Illustrator Alan Barnett's pictorial text undermines the historical place of women and war in so many ways. While Lottie's concerned and somewhat sorrowful appearance on the book's cover is arresting and the sadness of her story is reinforced through the use of blurred watercolor and a single cross at the base of which poppies grow, throughout the book the images often appear too modern and anachronistic. They certainly work to undermine the story of Le Gallais as a brave and independent woman. Instead, they replicate stereotypes about women and war.[43] The close connection to a "real" story and the ready accessibility of images about women, the First World War, and especially Charlotte Le Gallais herself invite questions as to why an illustrator would deviate so much from available historical material. Lottie is portrayed as if she were in her late teens or early twenties, despite her actual age being in her early thirties. This may be an attempt to invite empathy from a young audience. As Peter Jachimiak has pointed out in his analysis of the British comic *Charley's War*, constructions of boyhood assist in bridging the generation gap in telling historical tales to a modern teenage audience.[44] Relying on tropes of girlhood rather than maturity in the depiction of women does not work in the same way, partly because there is not the same kind of seamlessness built into histories of women's participation in war as there is about boys becoming men through soldiering. The youthful characterization of Lottie also falls into the perpetual trap of portraying women as girls in need of protection and hinders the portrayal of nurses as experienced professionals, and thereby, further marginalizing their importance and contribution.

Throughout the book, the pictures are stylistically late twentieth century and often portray the nurses without giving any indication of the rigid discipline they adhered to. Nurses were only allowed to be out of uniform "after dinner," and they would certainly not have appeared on deck of the ship during a public occasion without wearing their capes and veils—quite unlike illustrator Alan Barnett's portrayal of the *Maheno* leaving Wellington where one nurse is standing on deck out of uniform and without a veil.[45] Photographs from the time of nurses on the decks of hospital ships still show them in veils and capes (see figure 6.1).[46] It would also have been highly unusual for the kind of free and easy mixing between nurses and men, soldiers, and hospital staff alike that is indicated in many of *Lottie's* pictures. Nurses were often segregated from soldiers and closely supervised in social situations, like the dances and fancy dress parades that took place on board the *Maheno*. It is also extremely unlikely that any nurse would have appeared on deck, regardless of the circumstances, with her hair unbound. Regulations pertaining to the professional appearance of nurses were strictly adhered to and helped to maintain the separation of nursing staff from the men.[47]

The separation of female nurses from male soldiers would have been extremely important for a nation that was reluctant to send its nurses to the battlefronts and was anxious about the virtue and sexual morality of soldiers, nurses, and civilians. The intimacy of the jobs required of nurses rendered their position precarious in terms of what was acceptable for a woman to know, see, and do at the turn of the twentieth century. This did not prevent the idolization of nurses as angels of mercy or their sexualization in the imaginations of men. There is a long history of the sexualization of the nursing profession that continues into the present.[48] No studies specific to New Zealand have been conducted, but it is safe to assume that some of the sexual stereotypes of nurses are as much a part of New Zealand attitudes as they are of Australian or British cultures. Katie Holmes, in her discussion of Australian nurses during the First World War, states that these women are separated from the legends of Australian heroism in war because, while they were necessary, they posed a potential sexual distraction and because their presence was a reminder of the fragility of the male body and its helplessness when wounded.[49] The imagery of nursing is complicated, drawing as it does on notions of sacrifice and devotion, while the sexual threat these women might pose if their veils were ever removed went largely unspoken but not forgotten in Australian and British traditions.[50] The sexualization of nursing as a profession is still present in society, with the "naughty nurse" a staple of pornographic and popular culture. Barnett taps into the sexualized stereotypes of nursing by depicting Lottie and her colleagues as young and carefree in their dress and interaction with male medical and military personnel. Even if the youthfulness of the nurses is an acceptable device to assist in telling Le Gallais' story to children, the sexualization of these women is more difficult to reconcile. The determination, discipline, and hard work of the NZANS nurses are not replicated in Barnett's representations of them, with flowing locks, whose behavior is far less professional than it is titillating. On one page, Lottie is depicted looking straight at the reader in a most suggestive manner while she stands in a storage room with a doctor who appears to be in a laughing and flirtatious conversation with her.[51] The familiarity of the conversation supposedly references the social interaction mentioned in the written text on the adjoining page but more immediately taps into discourses of femininity and loose sexual morality.[52] In a similar way, the wholesome fun of devising fancy dress costumes described in Le Gallais' correspondence and referenced by Lockyer is depicted by Barnett in an anachronistic way that has little to do with the descriptions in the written text.[53] Instead of doctors "dressed up as nurses" and a nurse using sheets, boxes, and electric torches to depict the *Maheno* as described by Le Gallais, Barnett has the passengers socializing with painted faces and polished costumes that would look more at home in a book about Studio 54, the famed 1970s New York nightclub. Rather than a "fancy dress parade," as described in Le Gallais' letters and the official *Maheno* booklet published in 1916,[54] the book pictorially presents what looks like close fraternization at a party—public familiarities that would not only have been frowned upon but are highly unlikely to have occurred.

No doubt *Lottie* can be read as a "story book" where the basic story can be understood without direct reference to the pictures that accompany the written text.[55] However, as Christina Desai points out, "illustrations undoubtedly color readers' reactions to the story. Art and text are inextricably linked to create meaning that could not be communicated in any other way."[56] If the purpose of this book is to inform young people about lives and events a century ago—it can be assumed that this audience has little background knowledge of this history—then surely the illustrations of such an educational text should reflect its subject matter? The pictures should certainly not be treated with the attitude that they are "just pictures" for what is "only" a children's book. This undermines both the respect that the audience of the story deserves and the respect supposed to be owed to the significance of the First World War. A big difficulty for scholars, critics, and writers of texts aimed at young audiences is that while children's literature is deemed an essential part of the formation of a child's worldview, it is often, contradictorily, also not treated as "real" literature precisely because it is aimed at children. The impulse to censor the images children are exposed to often competes with the idea that children are not mature enough to grasp the complicated language of images or see the connections between different kinds of representations. It is important not to underestimate the capacity of children to read images and text. If children continue to be exposed to books that depict women in a frivolous and sexualized way, even if they do not understand exactly that this is what has occurred, then the contributions of women will continue to be marginalized in the public imagination.

This is not to say there are no picture books where authors, artists, and illustrators have collaborated and the written and pictorial texts appear to be telling a different story. For example, the subtle signs in Tony Kushner and Maurice Sendak's recent *Brundibar,* which, while telling the story of two young children trying to get milk for their sick mother, also allegorizes the experiences of Jewish children at the Theresienstadt concentration camp.[57] But in the case of *Lottie*, it is not unreasonable to expect the "faction" to be closer to the available archival pictorial material and widespread knowledge of images from the period than the "fictional" pictorial representation posited here. *Lottie* is, after all, presented by the publisher, author, and educational institutions as historical rather than allegorical in both form and function. The lack of care taken with these illustrations detracts from the importance of telling noncombatant stories and the role of the nurses who worked so hard under the most appalling conditions. Whether intentional or not, the work of these women has been trivialized here, just as it was in 1916 when the bulk of their activities were largely absent from the official *Maheno* souvenir publication.[58]

Children's literature about war from the late twentieth and early twenty-first centuries often functions with two aims in mind: a commemorative "lest we forget" function and an education function so that the atrocities of the past may never happen again.[59] This is particularly true of Holocaust narratives, but the picture books centering on New Zealand and the First World War operate in similar ways. The phrase "lest we forget" alone is extremely problematic. What is remembered in New Zealand is more often a mythology—what supposedly happened—than a real understanding of a century-old war, its participants, and its legacies. This mythology is extended through memorialization in the classroom with wreaths and poppies on ANZAC Day but with little or no explanation of their symbolism.

It is further heightened by the use of fictional and "factional" texts written and published in the late twentieth and early twenty-first centuries without an appropriate contextual explanation of the kinds of texts that they are and the past they represent. This is a part of the trend that is evident in a school curriculum that teaches the poetry of Wilfred Owen and Siegfried Sassoon as if it was representative of every soldier's experiences. Just as learning the "legend" supplants learning the problems of early twentieth-century history, many of the Great War's participants who were not combatants, even when they were casualties, like the nurses who died on the *Marquette*, are marginalized. The books aimed at young people in New Zealand are often a part of this pattern of active forgetting. The fragmented stories conveyed in commemorations, school curricula, and, of course, picture books must be more confusing than illuminating to children, who in general have no direct experience of war and for whom this war is beyond living and spoken memory.

The exclusion of women's experiences from war narratives, in general, and ANZAC mythologies, in particular, has been well commented on both in Australia and in New Zealand. On the surface, the publication of *Lottie: Gallipoli Nurse* represents a step in the right direction in filling some of these lacunae. However, it must be questioned whether this was the intent when it is a soldier who is the real main character of the book and not a nurse at all. The fragmentary, intertexually dependent *Lottie: Gallipoli Nurse* acquires its meaning from the weight of its archival origins, connections to museum exhibits, and the national mythologies of ANZAC. Without the archival material on display in the museum, *Lottie* may very well have no contextual anchor at all. If the framework of Charlotte Le Gallais' story is dismantled, the book merely reiterates the clichés and body counts that are a part of ANZAC mythology but *not* that of a nuanced history of New Zealand at war. The focus on the *Gallipoli* rather than *Nurse* part of its title and its conflation of the fictional *Lottie* with the real Charlotte Le Gallais through word and image only widen the distance between memorializing and understanding the past. If contexts are misunderstood or misaligned and skewed toward soldiers even in the narratives of those who did not, or could not, fight, how can future wars be prevented? How can we prevent the noncombatant children of today from becoming the combatant soldiers of tomorrow if the stories they are presented with replicate rather than interrogate these mythologies? How can the understanding of alternative narratives be fostered when the texts being offered to children contain factual inaccuracies and illustrations that trivialize their subject matter? I fear that the production of such texts form part of the vision blurred through tears that is the public memorialization of a war that is never remembered in its entirety but which refuses to be forgotten.

Notes

* The research for this article was conducted with financial assistance from the New Zealand Federation of Graduate Women, Auckland Branch.

1. John Lockyer and Alan Barnett (illustrator), *Lottie: Gallipoli Nurse.* Auckland, 1998 (2003).
2. Patricia Stroud, *Caesar the Anzac Dog.* Auckland, 2003.
3. Jock Phillips, Nicholas Boyack, and E. P. Malone, "Introduction," in Jock Phillips, Nicholas Boyack, E.P. Malone, eds., *The Great Adventure: New Zealand Soldiers Describe the First World War.* Wellington, 1988, p. 1.
4. For example, Martin Kay, "PM Wants More Prominence Given to Passchendaele," *The Dominion Post.* 5 October 2007, http://www.dominion.co.nz/4225960a6479. html. Accessed November 2007.
5. Deborah Montgomerie, "Reconnaissance: Twentieth Century New Zealand War History at Century's Turn," *New Zealand Journal of History.* 37, 1, 2003, p. 62.
6. Ibid., 74.
7. Jennifer Beck and Robyn Belton, *The Bantam and the Soldier.* Auckland, 1996.
8. John Lockyer, *Harry and the Anzac Poppy.* Auckland, 1997.
9. Chris Pugsley and John Lockyer, *The Anzacs at Gallipoli.* Auckland, 1999 (2003).
10. Stroud, *Caesar the Anzac Dog*; Glyn Harper and Bruce Potter, *The Donkey Man.* Auckland, 2004.
11. Jennifer Beck and Lindy Fisher, *A Present from the Past* (Auckland, 2006).
12. For example, the historian Glyn Harper has recently called for more military history to be taught in schools. Glyn Harper, interview by Kerry Woodham, NewsTalkZB, 30 September 2007.
13. *The New Zealand Hospital Ship "Maheno": First Voyage July 1915, to January 1916.* Christchurch, 1916, p. 59. Charlotte Le Gallais' correspondence says she was one of ten staff nurses to sail on the *Maheno,* but the publication above, published in 1916, lists her name as one of 14. It is possible that four more nurses were added to this contingent between Charlotte sending her letter to Leddra on 7 July 1915 and the ship sailing on 11 July.
14. See Sherayl Kendall and David Corbett, *New Zealand Military Nursing: A History of the Royal New Zealand Nursing Corps Boer War to Present Day.* Auckland, 1990.
15. Le Gallais Family. Papers. Auckland War Memorial Museum Library. MS 95/11, Folder 1 (AWMM).
16. Lockyer and Barnett, *Lottie,* p. 32.
17. Ibid. There has been an editorial mistake here further confusing this work of "faction" in the misspelling of Leddra Le Gallais' name— it should be Le*dd*ra not Le*d*ra.
18. The completeness of the Le Gallais family archive is one of the reasons why Charlotte, Leddra, and Owen (but most particularly Charlotte) have such a

prominent presence in the Auckland War Memorial Museum exhibit on the First World War. The museum file note says, "The collection will have immediate display use—it will probably form the focus for a segment of the World War I Gallery." Le Gallais Papers—Justification (Peter Hughes). Le Gallais Family. Papers. AWMM, MS 95/11.

19. Lockyer referred to *Lottie* as a work of "faction" in a personal communication with me, 30 March 2005.

20. Antoine de Baecque, *The Body Politic: The Corporeal Metaphor in Revolutionary France 1770–1800.* Stanford, 1997, p. 16.

21. Peter Hughes Jachimiak, "'Woolly Bears and Toffee Apples': History, Memory and Masculinity in *Charley's War,*" *The Lion and the Unicorn.* 31, 2007, p. 166.

22. Lockyer and Barnett, *Lottie,* p. 16.

23. Ibid., p. 9.

24. For example, A. A. Grace wrote in *The New Zealand Herald* on 1 August 1914: "The average New Zealander . . . especially the young New Zealander who lives in the country is half a soldier before he is enrolled. He is physically strong, intellectually keen, anxious to be led through being what he is, he will not brook being driven a single inch. Quick to learn his drill, easily adapted to the conditions of life in camp since camping usually is his pastime and very loyal to his leaders when those leaders know their job." Cited by Christopher Pugsley, *On the Fringe of Hell: New Zealanders and Military Discipline in the First World War.* Auckland, 1991, p. 9.

25. For Canadian and Australian examples, see Linda Cranfield and Janet Wilson (illustrator), *In Flanders Fields: The Story of the Poem by John McRae.* Toronto, 1996; Norman Jorgensen and Brian Harrison Lever, *In Flanders Fields.* Vancouver, 2003. Also, Kate Agnew and Geoff Fox, *Children at War: From the First World War to the Gulf.* London and New York, 2001.

26. For example, Michael Morpurgo, *War Horse.* Kingswood, 1982; Carol Fox, "What the Children's Literature of War is Telling Children," *Reading.* November 1999, pp. 128–129.

27. There is a long tradition of representing women who find themselves in unconventional situations as exceptional or different from other women, and this is present in media depictions, historical writing, and popular culture. See Sara Buttsworth, *Body Count: Gender and Soldier Identity in Australia and the United States.* Saarbrücken, Germany, 2007, especially chap. 5. Also see, Linda Grant De Pauw, *Battle Cries and Lullabies: Women in War from Prehistory to the Present.* Oklahoma, 1988. Feminist scholars have also commented on the lack of depiction of collective action, identity, or agency in discussions about women: Moira Gatens, "Corporeal Representation in/and the Body Politic," in K. Conboy, N. Medina, S. Stanbury, eds., *Writing on the Body: Female Embodiment and Feminist Theory.* New York, 1997, pp. 80–89.

28. Angela McRobbie, "Notes on Postfeminism and Popular Culture: Bridget Jones and the New Gender Regime," in Anita Harris, ed., *All About the Girl: Culture, Power and Identity.* London and New York, 2004, pp. 3–14.

29. Papers of John Duder, AWMM, MS 1160.
30. See, for example, Elizabeth Bell, Lynda Haas, and Laura Sells, eds., *From Mouse to Mermaid: The Politics of Film, Gender and Culture.* Bloomington and Indianapolis, 1995, pp. 4–6. David Rosen and Ismee Tames both discuss the problems of the construction of "childhood innocence" in their chapters in this volume.
31. The conflation of ANZAC, which primarily evokes an association with Gallipoli, and the poppy, which is associated with the battles in France, are interesting but not out of step with the ways in which the First World War is memorialized in New Zealand. Artificial red poppies are sold in New Zealand on ANZAC Day, whereas in Britain and Australia they are sold on "Poppy Day" (Armistice Day) on 11 November.
32. Lockyer, *Harry and the Anzac Poppy,* p. 12.
33. Julie Wheelwright, *Amazons and Military Maids.* London and San Francisco, 1994; De Pauw, *Battle Cries and Lullabies.*
34. With regard to young adult literature and the Second World War, see Ian Serraillier, *The Silver Sword.* London, 1956; Judith Kerr, *When Hitler Stole Pink Rabbit.* London, 1971.
35. Lockyer and Barnett, *Lottie,* p. 13.
36. Duder, Diary entry, 4 August 1915.
37. Charlotte Le Gallais, letter, 17 November 1915.
38. New Zealand Ministry of Education and Unitec, *Gallipoli Webquest, Social Studies online.* http://socialstudies.unitecnology.ac.nz/gallipoli_webquest/index.htm. Accessed November 2007; K. Catran, *Letters from the Coffin Trenches.* Auckland, 2002.
39. Anna Rogers, *While You're Away: New Zealand Nurses at War 1899–1948.* Auckland, 2003, pp. 77, 82.
40. Rogers discusses the sinking of the *Marquette* in chapter 6 of her book. She then cites the Royal George sinking as described by Lockyer in a later chapter (Rogers, *While You're Away,* p. 166).
41. For a recent example of this, see Gary Sheffield, "Britain and the Empire at War 1914–1918: Reflections on a Forgotten Victory," in John Crawford and Ian McGibbon, eds., *New Zealand's Great War: New Zealand, the Allies and the First World War.* Auckland, 2006, pp. 30–68.
42. Christina Twomey, "Australian Nurse POWs: Gender, War and Captivity," *Australian Historical Studies.* 124, 2004, pp. 255–274.
43. The anachronisms and inaccuracies do not only apply to the nurses. On page 22 in a picture, of what I assume is a nurse about to take a tourist ride on a camel near the pyramids, is someone swinging a golf club. While I cannot say with complete certainty that people did not play golf in the desert in 1915, this seems extremely incongruous, as does the depiction of the woman I assume to be a nurse wearing blue rather than a uniform. And it adds nothing to the telling of the story of Charlotte Le Gallais.
44. Jachimiak, "Woolly Bears and Toffee Apples," p. 169.
45. Lockyer and Barnett, *Lottie,* p. 4.

46. Kendall and Corbett, *New Zealand Military Nursing,* p. 60.

47. Ibid., chap. 6.

48. Terry Ferns and Irena Chojnacka, "Angels and Swingers, Matrons and Sinners: Nursing Stereotypes," *British Journal of Nursing.* 14, 2005, pp. 1028–1033.

49. Katie Holmes, "Day Mothers and Night Sisters: World War I Nurses and Sexuality," in Joy Damousi, Marilyn Lake, eds., *Gender and War: Australians at War in the Twentieth Century.* Cambridge, 1995, pp. 43–59.

50. For a discussion of this in English literature, see Catherine Judd, *Bedside Seductions: Nursing and the Victorian Imagination 1830–1880.* New York, 1998.

51. Lockyer and Barnett, *Lottie,* p. 7.

52. Lockyer's depiction of Lottie writing in a diary, rather than writing to a fiancé, may have indirectly assisted in the portrayal of a chaste and dedicated young woman. This is, however, undermined by Barnett's pictorial textualizations of Lottie and her fellow nurses.

53. Ibid., pp. 10–11.

54. *The New Zealand Hospital Ship,* p. 14.

55. Christina M. Desai, "Weaving Words and Pictures: Allen Say and the Art of Illustration," *The Lion and the Unicorn.* 28, 2004, p. 408.

56. Ibid., p. 408.

57. Tony Kushner and Maurice Sendak (illustrator), *Brundibar.* New York, 2003.

58. *The New Zealand Hospital Ship,* pp. 22–25 has a detailed discussion of the supervision of wards on the ship and the nurses are not mentioned once.

59. Kate Agnew and Geoff Fox, *Children at War: From the First World War to the Gulf.* London and New York, 2001, especially p. 138.

CHAPTER 7

"We Aren't Playing That Passive Role Any Longer": American Women's Protest of the Vietnam War

Penelope Adams Moon

On 16 March 1965, an elderly Quaker woman walked along Detroit sidewalks she had known for almost 25 years. As the Tuesday night traffic eased along, she stopped in front of a shopping center at the intersection of Grand River and Oakman Boulevard. For a good many of her 82 years, Alice Herz had been writing letters, begging her government to stop building nuclear weapons and, just as importantly, to stop threatening to use them. She was particularly distressed by the escalating war in Vietnam, a war most Americans were just starting to consider. But she was just an old Quaker woman; who was going to listen to her? As her hope in the efficacy of letters and petitions waned, she decided on one final act of protest. Frustrated, but at the same time hopeful, she doused herself with cleaning fluid and lit a match. Her body was immediately engulfed in flames. Despite the best attempts of passersby to snuff out the flames of what they would later describe as a "human torch," Alice Herz died a painful ten days later.

Herz was the first American to immolate herself as an act of protest against war. Long concerned with nuclear proliferation, she had been particularly affected by the self-immolation of Thich Quang Duc, a Buddhist monk, who had burned himself on the crowded streets of Saigon two years earlier. Quang Duc had chosen immolation to protest the abuses of Ngo Dinh Diem's regime in South Vietnam. With the help of the U.S. government, Diem had come to power in 1954 and had gradually become more paranoid and more hostile toward Saigon's non-Catholic residents. Quang Duc and a number of other Buddhists in Saigon burned themselves to call attention to Diem's repressive government and its American backers. In a letter she left behind, Herz identified herself with Quang Duc, denounced Lyndon Johnson for contributing to an arms build up, and implicitly condemned his manipulation of the facts to secure congressional approval for expanding U.S. military activities in Vietnam.[1]

Herz's protest was fairly exceptional among acts of American protest. Only seven other Americans, out of the hundreds of thousands that would eventually protest the war, chose immolation as their form of witness. Yet, Herz' act was also entirely typical in some ways. She was an American woman who chose to protest in a very personal and meaningful way in an environment that marginalized women's voices in all arenas, not the least of which was the movement against the Vietnam War. As in life, she received scant press coverage for her final protest act and history has veiled her importance, choosing instead to linger on the self-immolation of Norman Morrison, the first American man to immolate himself, seven months later.

In wishing to protest war, Herz faced an uphill battle. As the Vietnam War escalated in the 1960s, American women sickened by the war encountered an antiwar movement that had galvanized around the issue of the draft. Although a handful of intellectuals, policy experts, and religious activists had prophetically begun to protest American involvement in Vietnam in the early 1960s, the American antiwar movement flared to life in 1965 as President Johnson increased draft calls. Suddenly, Vietnam mattered to many more people who now faced the very real and terrifying possibility that they or someone they loved might be shipped off to the jungles of Southeast Asia. As news reports about the brutality of the fighting, the futility of American strategy, and the questionable morality of American tactics began to filter back home, more and more young men and their families began questioning the legitimacy of the war and the justice of the draft. As a result, the draft became the axis around which the antiwar movement rotated. To demonstrate their opposition to the draft and the war, hundreds of young men began turning in or burning their draft cards, refusing induction, fleeing to Canada or Sweden, or opting for prison.

With the antiwar movement so focused on the issue of the draft, antiwar women confronted the decision of how best to register their discontent and bring about the war's end. Because of immutable characteristics of biology and persistent sexism in the American Congress, women were, and would be for the foreseeable future, defined as noncombatants. Assumptions about the connections between biology, behavior, and personality also marginalized women in the political arena. Women faced the challenge of participating in both a political environment and an antiwar movement that privileged those eligible for combat service. While some might think that avoiding military service was a privilege, this "privilege" came at the cost of women's political influence. Without the prospect of combat service, women found it nearly impossible to be taken seriously as political actors, both among political powerbrokers and antiwar activists.

So how did women make their opposition to the war in Vietnam heard in a political and activist environment that marginalized their ideas and muffled their voices? As the historian Joel P. Rhodes has noted, "More often than not, in the realm of protest women have relied on ingenious and covert tactics for mitigating against patriarchy and a misogynist culture."[2] Gender frequently became the organizing principle around which women tackled the task of protesting the war. Said another way, American women drew upon their identities as women and their often shifting understandings of womanhood to protest the war in Vietnam. Women protested as mothers, wives, housekeepers, sisters, consumers, revolutionaries, race women, and more because their sex prevented them from speaking in purely political terms or from protesting as potential combatants, an identity marker crucial to the antiwar movement. In their protest, some women chose to reference traditional understandings of womanhood, what the historian Amy Schneidhorst labels as "pre-feminist gender norms," while other women negotiated gendered identities with emerging race- and class-consciousnesses to articulate their opposition to the war.[3] In each case, women faced the task of either transcending or manipulating existing gender standards to make their noncombatant perspectives heard in a cultural milieu that gave pride of place to political speech and combatant status.

This is not to say that the arguments and protest methods women chose were unique to women. Certainly, self-immolation was not an exclusively "female" protest tactic. My intent here is not to suggest that women always protested differently than men, but rather that women attempted to find ways to engage the war in Vietnam that were relevant to them and that were as effective as possible. Since identity is complex—one is not simply a woman but might also be black, wealthy, gay, urban, socialist, or Baptist—it would be misleading to suggest that gender alone dictated women's responses to the war. Having said that, though, it is significant that wider societal assumptions about sex did often compel women to protest in certain ways and led them more frequently toward certain arguments.

Braving the Masculinist Antiwar Movement

Women who hoped to protest the Vietnam War faced at least two gendered obstacles. The antiwar movement focused primarily on the draft and, thus, on draft resistance, a reality that rendered women, who were automatically noncombatants, relatively unimportant. But the draft-centric focus of the movement against the Vietnam War also transformed the tenor of antiwar protest itself. Newer antiwar activists, draft resisters, even pacifists, many of whom embraced the philosophy of nonviolence, began to cast their resistance in masculinist terms that rendered the culture of the antiwar movement very similar to what Marian Mollin described as "the warrior culture of military men."[4] Refusing the draft—that is, refusing to fight—carried with it the potential for emasculation. Since gendered assumptions connected masculinity with aggression, physical strength, and conquest, soldiering, or at the very least relishing conflict, was a key component of Western masculinity.[5] In announcing their refusal to soldier, draft resisters faced the real possibility of being labeled cowards. In effect, they risked being associated less with men than with women, whom gendered assumptions painted as weak, vulnerable, nurturing, and passive.

To minimize this potential, male draft resisters often constructed an androcentric and highly aggressive "resistance mystique," which emphasized "[m]anhood, manliness, [and] virility." This "unspoken agenda of masculinity," Mollin argues, "ultimately created the most formidable obstacles for women" in the resistance movement.[6] As the sociologist Barrie Thorne has contended, the resistance mystique "involved a highly sexualized, objectified definition of women (women, in this rhetoric, were usually referred to as 'chicks'). The presence of women, defined as girlfriends, admirers, and bedpartners, was used to buttress an almost swaggering masculine role."[7] In effect, rendering women powerless within the resistance allowed resistance men to salvage their identities as men. Pushing this further, sexism within the antiwar movement might have actually helped resistance men maintain their political clout in the public political arena.

This created a resistance environment that was not all that woman friendly, making it very difficult for women to fully participate in protest. The story of women's subordination within the New Left, civil rights, and antiwar movements is fairly well known as is the connection between the sexist experiences of women in these movements and the emergence of second wave feminism in the late 1960s.[8] But as Thorne has argued, the marginalization of women within the resistance was worse than that women experienced in either the student movement or civil rights movement because the centrality of the draft "more explicitly distinguished male from female participants and excluded women, even theoretically, from full participation."[9] Anne Weills, a participant in the Bay Area antiwar movement, remembered being frustrated at antiwar meetings. "Even if [a woman] said it well, half the time people would ignore you. Invisibility. That's what was so painful."[10] Jane Kennedy, a Catholic woman who felt compelled to speak out against the Vietnam War, remembered being disappointed by the gendered posturing surrounding plans to raid a local draft board. She characterized the arguing of some men as "a lot of bluster and bluff" that functioned to keep hidden men's "own feelings of inadequacy and their own spots of vulnerability."[11]

If this "bluster and bluff" kept men from exposing their vulnerability, it also kept women subordinated in the antiwar movement. As the sociologist Jo Freeman explained, "Men could resist the draft; women could only counsel resistance."[12] But by 1967, counseling resistance was not enough. The antiwar movement focused much of its energy on crippling the draft system through active protest, emblematized by the slogan, "From dissent to resistance." Draft-age men publicly burned their draft cards to comment on the injustice of the draft itself and to signal their unwillingness to fight in Vietnam. In one of the largest and most well-organized acts of resistance, some three dozen men turned in or burned their draft cards at Boston's Arlington Street Church on 16 October 1967. One of those that burned a draft card was the Reverend Nan Stone, a Methodist minister attending Boston University's School of Theology. Stone had convinced her friend, Steve Pailet, to allow her to burn his draft card but met with opposition from other protest planners. Although a seasoned activist, Stone had to work hard to be allowed to fully participate in activities planned by the New England Resistance. She continually ran up against gendered assumptions that left women with the domestic chores of resistance communities—what many routinely referred to as "shitwork"—while men planned and carried out risky protest actions. As Stone recalled, "I was never invited in to the inner circles, I had to push my way in."[13]

Besides burning draft cards, the antiwar movement was increasingly turning to even more radical actions that focused on the draft. By 1967, resisters had begun to use nonviolent direct action to physically disrupt the Selective Service System. This usually took the form of raiding draft offices and destroying actual draft files, inhibiting the government's ability to call up draftees. The 1968 Catonsville Nine raid in Maryland typified the new tactic. In that action, nine Catholic activists, including priests Daniel and Phillip Berrigan, carried hundreds of draft files out of the Catonsville draft office and burned them with homemade napalm. The raid itself would baptize the Catholic Left and catapult the brothers Berrigan to international notoriety. As their biographers and friends pointed out, the Berrigans became the desperadoes—a very masculinist term—of the Catholic Left.[14] Although two women, Marjorie Melville and Mary Moylan, participated in the Catonsville raid, they found it difficult to assume leadership positions. As one historian has argued, "the masculinist behavior and rhetoric [of the secular antiwar movement] dovetailed with the patriarchal and hierarchical underpinnings of the Catholic Church" to create a Catholic resistance community in which men "were cast as the true resisters" and emerged as resistance leaders.[15]

As Marjorie Melville and Mary Moylan's participation in the Catonsville Nine raid makes clear, some women eagerly participated in the destruction of draft files, which was very much against the law and almost always prosecuted. Draft-board raiding became a way that noncombatants could show how seriously they opposed the war. To avoid the sexism they encountered in many resistance communities, though, some women chose to develop their own resistance actions. Often this simply took the form of helping to raise awareness about the draft among draft-eligible men. The Berkeley group Campus Women for Peace, for example, distributed a leaflet that encouraged men to pledge to oppose the draft, an act that became prosecutable after passage of the Military Selective Service Act of 1967.[16] Other women tapped into the draft resistance movement by signing resistance statements, helping to gather and turn in draft cards, offering sanctuary to resisters, or engaging in disruptive acts during the public trials of male resisters. For example, two women calling themselves "Women Too" disrupted the 1969 trial of a draft resister by pouring red paint on the courtroom floor and pouring red dye on the steps of the federal building in which the trial was taking place.[17] Their decision to name themselves "Women Too" is telling. While the name played on the tradition of identifying draft resisters by the numbers of people involved in the action, it also implicitly referred to women's marginal political status. One can almost hear the women shouting, "Yes! We're angry, *too*! The war impacts us, *too*!" The name points to how the default identity of the antiwar protester in 1960s America was male.

Even when women planned resistance actions by themselves, society seemed to want to recast the action in traditionally gendered terms. In July 1969, five women calling themselves "Women Against Daddy Warbucks" spent about three hours slicing through draft files, tearing out phone lines, and disabling the typewriters in a midtown Manhattan Selective Service office. When they were finished, they exited the building and began throwing the cut-up files up in the air in an almost celebratory fashion. In all, they managed to destroy about 6,500 1-A high-priority draft files.

Yet such was the public's identification of draft resistance with men that even the reporting of the raid by Women Against Daddy Warbucks cast men and women in traditionally gendered roles. Despite having been planned and executed by women, the *New York Times'* reporting on the raid chose to highlight the scuffle between nearby male antiwar activists and law enforcement officers. The *Times* described how "some men war resisters . . . attacked the Federal agents with their fists." In the ensuing battle, both male protesters and male law enforcement agents tried to protect the females involved. "The F.B.I. men rushed their prisoners as quickly as possible to cover . . . [as] the agents around one pair [of female prisoners] beat off their attackers as they moved through the crowd." Later male protesters attempted to "free [Kathy] Czarnik," creating a narrative that cast men as liberators and women as victims in need of rescue.[18]

Women Against Daddy Warbucks had staged the action, in part, out of frustration with sexism in the larger antiwar movement and, in the words of participant Maggie Geddes, because the "draft affects men and women." The women wanted to publicize how women helped to sustain the draft through their work in draft offices and their passivity about the "corporate military machinery." But as Geddes asserted, "we aren't playing that passive role any longer."[19] Yet, as the *Times*' reporting made clear, few gave the women the credit for planning the raid by themselves. Jill Boskey, another member of Women Against Daddy Warbucks, remembered that when she was first arrested, the FBI agents tried to get her to name the men she was fronting for, and even men in the movement kidded the women about not having planned the action.[20]

As a result, some women draft resisters chose to engage in resistance activities in ways that actually emphasized their female identity; they decided to capitalize upon existing gender systems to protest the draft. For example, some women posed as sisters of inductees or as secretaries to gain access to military bases and disrupt the induction process. While male draft resisters could pose as draft registrants or preinductees, women's noncombatant status made them conspicuous and visibly out of place among potential soldiers. Ironically, antiwar women realized that their relegation to supportive or appendage roles in the American workforce and their status as noncombatants could actually work in their favor. The "invisibility" that frustrated some women in resistance meetings enabled other women to slip onto military grounds unnoticed or unquestioned, putting them in a position to disrupt the induction process and engage in consciousness-raising with draft registrants.[21]

Women's sexual identities also frequently factored into their resistance. Existing gender systems simultaneously cast women as both asexual and potential objects of sexual conquest. To lure draft registrants or even active-duty GIs into draft-resistance environments, resistance organizations often emphasized the presence of women at resistance-planned parties and events. In one instance, the New England Resistance publicized that it was planning a "huge, incredibly noisy, chick-laden" party, while another leaflet enticed GIs to an event with "beer and chicks and things."[22] Women, however, were not always passive players in the sexualizing of their roles in the antiwar movement. In perhaps the most famous example of women sexualizing their own participation in the resistance, many women accepted the slogan "Girls Say 'Yes' to Guys that Say 'No,'" although, as Barrie Thorne points out, many did so "with jest and ambivalence."[23]

Maternalist Responses

The "girls" that said "yes" to the guys that said "no" were playing up aspects of existing gender systems that cast men as highly sexualized beings—so sexualized, in fact, that they would risk imprisonment for draft evasion to potentially have sex. Indeed, capitalizing upon the assumption of gender difference was one of the most powerful and effective ways that women could challenge the war. Because their noncombatant status frequently prohibited them from fully participating in draft-centric protest, women were often forced to either accept a supportive role in protest or emphasize aspects of their identity that made them different from men. However, this was not always a fallback decision. Unlike the gendered role-playing involved in gaining access to military bases during the induction process, some women were explicit that their gender-specific identities compelled them to protest the Vietnam War. Stressing the bonds of sisterhood and the shared experience of motherhood, women banded together to chastise male war-makers and argue against the war. It was in their very roles as female noncombatants that many women found some of their most salient forms of protest.

Sometimes the sororal and maternalist themes could be subtle. In their 1965 statement "For a Ceasefire in Vietnam," the Berkeley Campus Women for Peace condemned the use of biological and chemical weaponry and the war's oppression of peasants, indiscriminate killing, political corruption, excessive cost, and nuclear potential. Yet, to make these arguments, which were common political critiques, the Campus Women for Peace used the testimony of Mai Thi Chu, a Vietnamese woman. The Campus Women for Peace's statement quoted Chu, who testified that she had seen how the use of chemical and biological weapons had affected women, children, and old people. "I have seen children with swollen faces and bodies, covered by burns," Chu reported. "I have met women blinded or suffering from sanguinolent diarrhoea."[24]

This statement is complex. It projects a gender-neutral position as a relatively straightforward political critique of the war, no different from statements that might have been issued by male-dominated organizations like Students for a Democratic Society or the Committee for a Sane Nuclear Policy.[25]

Yet gendered assumptions are woven throughout the statement. We are meant to be shocked that women and children—that is, innocents—are dying in Vietnam. Mai Thi Chu's testimony reifies the idea that women, like children, are helpless victims in warfare and that killing them is more barbarous than killing men. At the same time, though, the statement challenges existing gender systems in privileging a female voice. The Berkeley Campus Women for Peace seemed to want to both empower women and galvanize them to action by helping them identify with Vietnamese women as women—as sisters and fellow mothers. Whether embracing or challenging gendered assumptions, though, the subtext underlying the statement is that war is not something that women do—they suffer as innocents or they speak out against it.

Other women pushed this assumption in very explicit ways, claiming that there was a universal difference between men and women and that difference lay in women's alleged natural aversion to war. Valerie Clubb, a member of Missoula Women for Peace, explained her antiwar activism in almost genetic terms. "Oh, I have had an instinctual feeling for peace, for peaceful relations between people, even without the fear of war and death and so forth."[26] May McDonald, another Missoula woman for peace, was convinced that if women around the world would just join peace groups, wars would cease to happen.[27] At a 1972 "Women's Hearing on the War in Vietnam" in St. Louis, panelists and audience members heard a number of women speak to specific issues related to the war. Testifying during the hearing, Eda Houwink argued that war was "exclusively a male activity." Men, Houwink argued, "declare the war, they plan the war, they pass the laws to finance it. No woman does this kind of thing. War is a game with men, by men, and for men."[28]

For many women, this instinctive aversion to war was rooted in their identities as mothers or, at the very least, as potential mothers. Many female protesters began their activism only after they became mothers. A number of the members of Missoula Women for Peace recalled coming to oppose the war out of concern for their own children. May McDonald remembered becoming "more and more alarmed [about the Vietnam War], especially as I had three sons who would likely—very, very likely—be inducted or drafted, and I thought it wasn't enough to condemn it verbally, I should do something about it, politically in every way that I possibly could."[29] This justification for opposing the Vietnam War—to protect one's sons—seems not to have been an argument many men offered publicly, perhaps because men were assumed to possess the higher-order thinking skills of logic and critical analysis; the assumption was that men would challenge the war politically (that is, rationally). Fear for one's children was an emotional and instinctive response to war suitable for the sex defined as being primarily emotional in nature. Moreover, the masculinity of fathers would to some extent have been tied up with the willingness of their sons to go off into battle. A son who eschewed his martial responsibilities indicted a father as a failure.

Maternalist protesters considered their identities as mothers key to stopping the war. "Being a mother," Anniece Allen, another witness at the Women's Hearing on the War in Vietnam, explained, "I am concerned with life in general." Although Allen was a psychiatric charge nurse by profession and might, therefore, have critiqued the war from a mental health perspective, she explained her opposition to the war in biological terms and sought to reach out to the female audience by emphasizing their shared roles as mothers. "Women are giving birth to children every day," Allen noted. "Are we giving birth to them, raising them, training them only for them to die?"[30] Eda Houwink continued this maternalist theme when she stated that women, because they "bear the children of the world . . . have a more intense feeling about being our brother's keepers [sic]. Men are much more extravagant with both their seed and other people's children."[31]

This maternalist approach to opposing the war is best exemplified by Women Strike for Peace (WSP), the phenomenally popular antiwar organization. Founded in 1961, WSP coalesced after thousands of women responded to a call by a group of Washington, D.C., women to publicly protest nuclear testing. Although its ranks included working women, WSP's public face was that of the housewife and mother. By 1966, WSP had come to focus on the situation in Vietnam, engaging in a number of protests conducted, according to the participant-historian Amy Swerdlow, "in the name of outraged motherhood."[32] Emblematic of WSP's maternalist or motherist agenda was the organization's slogan during its draft-resistance marches: "Not My Son, Not Your Son, Not Their Sons."

Unlike younger female activists interested in draft resistance, WSP-ers successfully engaged in draft resistance largely because they adhered to middle-class femininity during their protests. Describing a 1962 WSP protest, the *New York Times* reporter Jeanne Molli noted the presence of mothers, grandmothers, and "the more formidable appearance of mothers-in-law," with all the visual imagery created by such a turn of phrase.[33] In her study of the Chicago branch of WSP, Amy Schneidhorst noted that "women activists who projected a lady-like, middle-class motherly image felt they would be more readily accepted by the women they hoped to recruit and less likely to be attacked by officials."[34] Thus, WSP-ers marched dressed in coats and pantyhose and often carried nonthreatening (feminine?) symbols such as flowers, white doves, shopping bags printed with peace slogans, and even their own children. Women understood that by embracing existing gender roles, they were more likely to be effective in a political environment intolerant of traditional political activity by women. Rather than resisting the limitations placed on them as noncombatants, maternalist protesters embraced their noncombatant identities, emphasizing the very qualities that had rendered them noncombatants in the first place.

Maternalist protest, however, transcended simply emphasizing women's reproductive capabilities. After condemning men as natural war-makers, Eda Houwink challenged the female audience at the Women's Hearing on the War in Vietnam to commit to a kind of "global housekeeping" and change the social order to prevent war. The "global housekeeping" Houwink called for echoed themes drawn upon by maternalist protesters and reformers since at least the nineteenth century. Although historians have correctly demonstrated that the confinement of women to the private domestic realm inhibited their political participation, they have also recognized that women's power within the home increased as men assumed identities as public, political beings. Assigned the responsibilities of nurturing children and creating moral environments for families, women eventually entered the public fray on the grounds that society, like a disheveled house or an unruly child, needed a woman's moral authority and organizational skills.[35]

Maternalist protesters in 1960s America continued to use this line of reasoning, arguing that women were natural peacemakers not simply because they bore children but because they created and managed the home environment. As Florence Johnson of Missoula Women for Peace explained, although women might not face induction, they could help others to resist the draft through their domestic roles. "As mothers, as wives," Johnson argued, "in helping create a happy home life, we encourage compassion and concern for others," which, according to Johnson, could foster "sound mental health and world peace."[36] Lois Barrett, testifying during the Women's Hearing in St. Louis, chastised women for socializing sons and husbands to grow up "believing this myth about how they must wage war to protect their women and their homes."

> Let us raise our sons in such a way that they will not be the dirty workers for a few powerful men. Let us raise our sons in such a way that they will not go to foreign lands and rape and use the women there. . . . [Let] us redefine the meaning of saint, and courage, and power for our sons and daughters.[37]

Other maternalist protesters developed protest tactics that drew upon women's traditional roles as cooks. Along with writing letters to elected representatives and joining in area peace marches, the Missoula Women for

Peace held bake sales and rummage sales to educate the public about what was going on in Vietnam. Sandra Perrin remembers shrugging off her husband's exasperation to bake cookies for one of the organization's many antiwar bake sales. "[M]y husband was wondering at one point, 'Sandra, what are you doing? Another batch of cookies for the Missoula Women for Peace?'" But Perrin and other Missoula Women for Peace believed that the cookies enticed people to come over to the organization's table and peruse the group's antiwar materials, perhaps playing on the old adage "The way to a man's heart is through his stomach."[38]

Along with mothering and cooking, the standards of female domesticity charged women with the task of shopping for the family, a role consistent with women's alleged instinct for nurturing and their responsibility for keeping house. Even at the nadir of female domesticity, American women had always worked as household financial managers. In the strictly patriarchal antebellum South, white plantation mistresses were expected to manage the household budget and keep the cupboards well stocked.[39] In the immigrant households of the industrializing North, working children and husbands turned their checks over to the mother or wife. As the historian Elizabeth Ewen explained, immigrant women "had to secure . . . goods in the marketplace—and since the wife was the center of the household, it was her responsibility to manage the conversion of wages into necessities."[40] Women in 1960s America remained crucial to the consumer economy and played increasingly important roles in economic decision-making in the household.[41] Antiwar women recognized their economic roles as sources of power and sought to capitalize upon their consumer identity to protest the war in Vietnam.

One organization that neatly combined a maternalist and consumerist approach to protesting the Vietnam War was Another Mother for Peace (AMP). Founded in 1967 by a group of women in Beverly Hills, California, AMP adopted many of the same maternalist positions as WSP as demonstrated by its popular slogan "War is not healthy for children and other living things." As an AMP mailer explained, "We who have given life must be dedicated to preserving it."[42]

Yet AMP pushed beyond appealing to women as mothers to appeal to them as consumers. AMP sought to encourage women to use their power as consumers to challenge the war in Vietnam in two ways. Unlike WSP, which was not centrally coordinated and focused primarily on marching, draft counseling, and acts of civil disobedience, AMP centered its activities on raising awareness through the sale of posters, patches, jewelry, pins, tie tacks, stationery, and peace cards emblazoned with its slogan. Although, theoretically, anyone could buy these items, AMP literature targeted women, using phrases such as "just in time for holiday gift-giving."[43] AMP sponsored card campaigns, which encouraged consumers to purchase boxes of cards to send to politicians in Washington, D.C. The organization's "Peace by Christmas" campaign, for example, planned to flood Washington with Christmas cards that urged elected officials to stop the war. In its Mother's Day campaign, AMP sold at least 20,000 cards that made their way to the Capitol.[44] Through their purchasing of peace paraphernalia, women could harass congressional war-makers and fund AMP's effort to stop the war.

Groups like AMP, though, also targeted women as consumers from a different angle. "If we are to believe the figures," Alberta Slavin argued during the Women's Hearing on the War in Vietnam, "women control and spend over 90% of the money in the economy. If women as consumers really want to end the war, we could conduct a devastating economic boycott." She encouraged women to connect with each other and to stop shopping.[45] Housewife Mary Jane Badenoch proclaimed that she was willing "to live on nothing but necessities, to buy no new labor-saving or entertainment products and to let the companies who contribute to the war know [that she would] no longer succumb to their advertising."[46] Ora Lee Malone condemned companies such as Whirlpool, Bulova Watch, Motorola, and General Motors for producing war-related materials. She understood that these companies valued women as consumers and felt women were in a position to compel them to help create a peaceful society. "As consumers," Malone pointed out, "we can be very influential."[47] One antiwar flyer pointed out that women's consumer identities made them responsible for the war but could also be the key to stopping it. "Women, though we don't carry guns, are as much a part of the war effort—and the society that makes wars like this possible—as are men." The flyer announced a protest in which women would condemn war profiteering and pledge to "stop paying for the destruction."[48]

In a concerted effort to raise awareness of how women's purchasing power fed the war machine and, conversely, could starve the war machine, AMP produced and distributed a film titled *You Don't Have to Buy the War, Mrs. Smith*. The 1970 film featured former Miss America and AMP member Bess Myerson Grant explaining how many consumer goods manufacturers were also crucial components of the defense industry. Along with the film, AMP sponsored a letter-writing campaign in 1970 to challenge consumer goods manufacturers to abandon their roles in the defense industry. Calling on women to band together as consumers, the June 1970 AMP newsletter encouraged women to write to company presidents and board members and, in effect, scare manufacturers away from defense work.[49]

While draft-card burning and draft-board raiding placed men at the center of the debate about the war, engaging the war from an economic perspective allowed women—a dominant force in the consumer economy—to speak with power. By addressing the war as consumers and as people who, often without their knowledge, helped equip the military, women could more fully participate in antiwar protest. Often, though, antiwar activists baulked at moving away from combatant-centered protest. Jane Kennedy, who initially supported draft-based protest actions, remembered her frustration with her fellow protesters:

> [W]e were talking about draft boards and I had made the point . . . that I could not see the value of staying just with draft boards, that we had to share the symbolic action, that it was not just draft boards that were culprits here. That the industrial people, those who made money out of things that soldiers used to kill other human beings with had to be coupled with draft boards to show the military-industrial complex kind of thing. That I could understand about the draft boards, but I simply had a wider horizon than that.[50]

As a noncombatant, Kennedy's participation in draft-board raiding implicitly carried less weight than that of a male (that is, a potential combatant). Eventually, Kennedy's group, which humorously adopted the name "the Beaver 55" (although there were only eight members), compromised and decided to protest in tandem the Selective Service and the military-industrial complex. In November 1969, the Beaver 55 entered the Dow Chemical plant in Midland, Michigan, a company that produced napalm, and scrambled magnetic tapes containing biological and chemical research. A week earlier, the group had successfully destroyed the draft records of 44 Indianapolis draft boards. The Beaver 55's decision to couple draft- and consumer-oriented protest reflects a recognition by the participants that the war was important to more than just those who served in the military. Noncombatants, among them women, were similarly responsible for the war's destruction and equally able to protest toward its end.[51]

Targeting Dow Chemical made a lot of sense to many women. Women were particularly angry about the American military's use of napalm in Vietnam and took the lead in public acts of protest against its manufacture and distribution. Many women recognized that the manufacture of napalm was intimately linked with their roles as consumers and homemakers. Donna Allen of WSP explained that the use of napalm galvanized women because of its horrifying impact, particularly on children. But WSP's Cora Weiss laid part of the blame on women themselves, who, through their roles as consumers, funded the production of napalm. As Weiss told the antiwar movement chronicler Tom Wells, "The guys who made napalm in Vietnam made something that you used in your kitchen everyday [Saran Wrap], so you could understand it. And what makes it stick to food is what makes it stick to babies." With this knowledge, some women focused their public protest not on draft resistance but on napalm resistance. In May 1966, four women successfully kept trucks from distributing napalm for seven hours. They continued their protest in Alviso, California, where they were arrested for preventing napalm bombs from being moved from storage facilities to barges. Their actions prompted the town of Alviso to prohibit the company from storing napalm.[52] As mothers, women were appalled by napalm's effects on children. As consumers, they knew they could do something about it.

Community Caretakers

Women did not simply exist as consumers of manufactured goods but as taxpayers and consumers of government services. Having helped to create the modern social welfare state in the late nineteenth and early twentieth centuries, women understood how important government programs were to the health of local communities and quickly recognized that the Vietnam War was crippling the government's ability to respond to people in need. This was a realization shared by Lyndon Johnson, who resented the way that the situation in Vietnam siphoned funding away from his domestic reform agenda. "I knew from the start," Johnson remarked, "that I was bound to be crucified either way I moved. If I left the woman I really loved—the Great Society—in order to get involved in that bitch of a war on the other side of the world, then I would lose everything at home."[53] Just as Johnson suspected, many American women understood that it was impossible to pay for both guns and butter. In "Women and the Draft," activists from the antiwar organization the New Mobe called upon women to "stop paying for the war—the money is needed at home for free child care, medical care, decent welfare payments, better education, and housing."[54] Catherine Stenger, a housewife, lamented that "over 50% of the [federal] budget is spent on war-related activities when thousands at home and abroad are starving."[55]

Women of color were acutely aware of the connection between insufficient social services, community suffering, and the war in Vietnam. Disproportionately represented among America's poor and majority residents in America's ghettoes, black women came to the conclusion that the government had the wrong budgetary priorities. Responding to a polling inquiry, Ms. Thomas, an African American secretary in Chicago, criticized

President Nixon for "spending billions of dollars of taxpayers' money in Vietnam . . . when this money could be used to help persons at home who are not working and need aid."[56] Black women facing the challenges of poverty resented the idea that communism was a threat worth spending billions to combat. "Why should we worry about Communism over [in Vietnam]," a black mother asked, "when we have poverty over here?"[57] Coretta Scott King, a vocal opponent of America's activities in Vietnam, drew very specific connections between poverty and the war. "As long as we kill men, women and children in Vietnam, millions of poor people face unnecessary death and suffering in America," King prophesied. "As long as we lay waste to the beautiful countryside and communities in Vietnam, we shall see destruction and chaos in the ugly ghettos of America."[58]

Chicanas were equally alarmed at how budgetary priorities in Washington left their communities without sufficient social services. Tanya Luna Mount, an activist who participated in the emerging Chicano/a movement in California, found it ridiculous that her community's schools waged a continual battle for funding. "Do you know why they has no money for us?" Mount asked, "Because of a war in Vietnam 10,000 miles away, that is killing Mexican American boys—and for WHAT? We can't read, but we can die! Why?"[59]

Beyond its impact on social programs, though, many women of color felt that the war was having a detrimental emotional impact on their communities. Jacqueline Mack, a black woman running for office in Missouri, testified before the members of the Women's Hearing on the War in Vietnam, that the war was destroying her people.

> There are thousands of women, and women without husbands, and children without fathers . . . and yet millions of dollars are spent for warfare. . . . This war has pursued every describable tension and bitterness among the people of my community. And we also feel that many other outbreaks of violence in our country is the direct result of built-up tension and emotions brought on by the Vietnam War. . . . We cry out in desperation that right here in America there are numerous wars going on: war on poverty, war on crime, war on unrest, war on inequality—and many other types of wars that could be fought and won with a quarter of the millions of dollars that are being spent in Vietnam.[60]

Like Mack, a number of other women of color chose to protest the war by speaking out, an act that could still be risky in the racially charged atmosphere of the 1960s. In one of the most notorious protests of the war, the singer and actress Eartha Kitt highlighted the connection between crumbling communities and the war in Vietnam. At the Women Doers'

Luncheon, hosted by Lady Bird Johnson in January 1968, Kitt used her invitation to speak out against the war. After listening to a program of speakers talk about how to ensure safe streets and battle crime in American cities, Kitt boldly informed the gathered women that they had absolutely no idea what they were talking about. While insufficient domestic spending contributed to urban problems, Kitt claimed that the draft was at the heart of juvenile delinquency. A black boy in America faced the prospect of being "snatched away from the mother and sent off to Vietnam," regardless of good grades or good behavior, Kitt informed the women. As a result, black youth turned to crime. Crime, she argued, was a way to avoid going to Vietnam.[61] Kitt reasoned that black youth rebelled against being drafted into a war that made no sense, and in the midst of this rebellion, black communities suffered.

Polls routinely showed black women to be the portion of the population most critical of the Vietnam War.[62] While whites might also come to antiwar protest out of frustration with the ways that the war limited social spending, black women's community worries often went beyond fiscal concerns. As one woman wrote in *Ebony* magazine, "We want—we demand—that all of you [black soldiers] come home alive and ready to fight poverty, hate and racism here in America."[63] Rosalynne Hughes, speaking during the Women's Hearing on the War in Vietnam, urged Americans to focus on "getting it together here . . . before getting it together some place else because it is definitely not together here." For Hughes, "getting it together" meant creating a truly equal society at home.[64] For many black women, caring for their communities meant eliminating racism and injustice. Many considered the Vietnam War to be a distraction from civil rights work and a symptom of persistent racism in the United States. Thus, for a lot of black women, antiwar work was race work, and race work was essential for community survival. "Women are going to have to move to stop this system from drafting their sons," the National Black Anti-war Anti-Draft Union proclaimed; "We have an obligation to our people first."[65]

The race consciousness that infused the arguments black women made against the Vietnam War frequently fostered a sense of solidarity among black activists with the people of Vietnam, whose situation, they felt, was analogous to their own. Early in 1967, Diane Bevel, wife of the civil rights activist James Bevel, working against the wishes of the State Department, became the first black woman to travel to North Vietnam and interview Ho Chi Minh. The beginning of Bevel's report on her visit stressed the suffering of women and children, echoing the maternalist concerns held by women in organizations like AMP. Bevel recounted how she had witnessed the "charred remains of an unborn infant blown from the body of [a]

Vietnamese mother." Indeed, Bevel admitted to first becoming interested in going to Vietnam after seeing pictures of a grieving Vietnamese mother holding a wounded or dead child. "I saw myself in this mother's place," Bevel explained. Bevel's maternalism, though, extended into a critique of what she characterized as the United States' genocidal war, a war she believed had "racial implications for the Vietnamese people and for black youths from America."[66] Like Bevel, Eslanda Good Robeson, wife of the longtime activist Paul Robeson and an activist in her own right, characterized the Vietnam War as "a war against a colored people engaged in a long valiant struggle for freedom and self-determination."[67]

Chicanas, too, cast the war as a racist imperialist war. In 1968, Betita Martinez helped found *El Grito del Norte,* a paper with a majority female staff that documented the Chicano/a movement. Martinez used the paper to educate her readership about Vietnam, often drawing explicit parallels between land conflicts in the United States and those occurring in third world nations like Vietnam.[68] Reflecting on her experience at an April 1971 women's conference on Indochina in the 5 June 1971 issue of *El Grito del Norte,* Dolores Varela remembered being struck by how the Indo-Chinese women she met had the same color skin as she did and how similar Vietnamese "concentration camps" were to American Indian reservations. The 29 August 1970 issue was even more explicit in linking the racial struggles of Chicano/as with those of the Vietnamese when it juxtaposed photographs of North Vietnamese children and *"campesinos"* (farmers or farmworkers) with similar images of children and workers in northern New Mexico.[69]

Mediated through their own experiences with racism and injustice, women of color often cast the war as one in which poor nonwhite communities struggled for dignity and survival. It is no accident, then, that women of color spoke vehemently against the war in community forums and from a community perspective. While such a local, often neighborhood, focus has tended to render women invisible to scholars of the antiwar movement, Kathleen Blee argues that we need to look beyond traditional spaces to hear women's voices and witness their protest. Women's activism, Blee notes, often "involves informal networks of friendship, kinship, or neighborhood rather than elections, positions, and hierarchies of organizations," and is "situated outside formal organizations: in the social networks of friendship or shared experience, the spatial commonalities of everyday life in homes, stores, neighborhoods, and schools, or the ties of sexual identities and attachments."[70] This is not to say that women did not protest using more traditional political forms; they did. But marginalized in traditional political arenas because of their sex and their noncombatant status, many women experienced the war and chose to engage it in terms that were familiar, closer to home, and vitally important to their lives.

Conclusion

On 17 June 1972, women from the St. Louis area gathered for what was billed as a Women's Hearing on the War in Vietnam. Over 40 women testified about the war and its impact on their lives and communities before a panel of 12 women. The women came from all walks of life—they were lawyers, churchwomen, nurses, housewives, professors, city council women, and wage workers. In her summary report on the hearing, the moderator Ann Q. Niederlander, an attorney from St. Louis, noted that the overwhelming sentiment of those in attendance was that the war was wrong. Niederlander's report highlighted women's concern with the economic dimensions of the war and its negative impact on the poor, on the health of the population in general, and on returning veterans, many of whom came home mentally ill and drug addicted.[71]

The Women's Hearing points to a number of ways that antiwar women engaged and confronted the Vietnam War. Sexism within the larger antiwar movement often pushed women toward sex-segregated forms of protest in which they could control the agenda and voice their dissent as they saw fit—the organizers of the Women's Hearing had created space for women's opinions. For some women, sex-segregated protests made sense as they were convinced that women's perspectives were inherently different from those men offered. An assumption of sisterhood and gender difference echoed throughout the testimony of many who spoke at the Women's Hearing. At the same time, although some of the women seemed to embrace a gender system that assumed difference, the format of the Women's Hearing also exuded confidence in women's ability to employ traditionally male forms of discourse—participants put the war on trial and acted as witnesses and judges against a male-constructed and male-enacted policy in Vietnam. While some women engaged the war from a maternalist perspective, others confronted the war in terms that did not highlight their female identity, using race, class, human rights, or economic arguments as the foundations for their critiques.

But the fact remains that women had a hard time being heard, whether in sex-segregated or mixed-sex environments. Their voices were muffled in part because of their noncombatant status. As feminist Kathie Sarachild explained, women's exclusion from the draft was not a blessing but "an emblem of . . . powerlessness compared to the men of their generation—as their 'No' to the war lacked the strength the men's had of being able to say

'We Won't Go'—and highlighted their more powerless and auxiliary position in the rest of society."[72] Like the frustration implicit in the name chosen by the women of Women Too, the sex-segregated nature of the Women's Hearing on the War in Vietnam on some level laid bare fundamental inequalities in society.

Although women's changing roles in the military and the frontless nature of current warfare have rendered women's status as noncombatants ambiguous, vestiges remain of the tradition of privileging combatant voices politically. In the U.S. presidential election of 2004, candidates chose to focus at length on the integrity of the military service of President George W. Bush and his challenger, Senator John Kerry, revealing how important military service is to those who hope to make their mark politically. As the candidates sparred over who did what and who avoided what, noncombatant Americans—that is, women—were pushed to the margins of the discussion.[73]

Nevertheless, what remains striking about the Women's Hearing on the War in Vietnam is that the women themselves felt competent and driven to comment upon the war, despite the fact that their noncombatant status rendered them peripheral in both formal policy-making and the antiwar movement. Even during an era when they could be definitively labeled noncombatants, they refused to let their noncombatant status silence them on the issue of the war. While their voices might have had little impact on ending the war, the protest of women during the Vietnam War helped bring to light the totality of war, that is, the pervasiveness of its impact. Although the war itself did not meet traditional definitions of a "total war"—at least not in the United States—women repeatedly exposed how war impacted a range of Americans. Noncombatant voices helped to bring to public attention the many ways—some subtle, some devastating—that they and their communities felt the war. The war robbed them of their sons, husbands, and brothers; it killed and maimed children; it made people sick; it reduced access to needed social services; it harmed the environment; it distracted Americans from other political battles; it caused tension and strife within communities; and it perpetuated a racist and sexist social order.

There is historical and political value to be gained from heeding noncombatants' perspectives on the broader consequences of war. Noncombatant voices have had a significant impact on the way war is waged. In the last half century, criticism of military tactics, much of it voiced by women, has prompted a more concerted effort to minimize "collateral damage" and protect noncombatants living in combat zones. The next step, though, is for noncombatant voices outside the (potential) combat zone to have a real impact on discussions about if, why, and when war should occur. Certainly the impact of war on combatants is dramatic and direct. But a policy-making process that focuses inordinately on "the troops," to the extent of marginalizing or demonizing noncombatant voices, though, is unsound, irresponsible, and undemocratic. A war's impact on those who do not, will not, or cannot fight is just as important as its impact on those closest to the combat. Taking into account the domestic impact of war, beyond the political implications for the party in power, would likely require a more careful, deliberate, and considered course of action. Such a process would acknowledge that war is a complex animal whose bite is felt not simply on the battlefield.

Notes

I would like to thank the faculty at Bethel College for their thoughtful insights and responses when I presented an early version of this paper during the 2006–2007 Faculty Seminar series. I am also indebted to Peter Miller and Jordan Penner, both students, for helping to edit the final version.

1. Alice Herz's suicide note, printed in Hayes B. Jacobs, "The Martyrdom of Alice Herz," in Shingo Shibata, ed., *Phoenix: Letters and Documents of Alice Herz.* Amsterdam, 1976, pp. 160–61. The *New York Times* reported that Herz's pocketbook contained a note that accused President Johnson of "trying to wipe out small nations," an implicit reference to Vietnam. David R. Jones, "Woman, 82, Sets Herself Afire in Street as Protest on Vietnam," *New York Times*, 18 March 1965.
2. Joel P. Rhodes, *The Voice of Violence: Performative Violence as Protest in the Vietnam Era.* Westport, CT, 2007, p. 162. Also see: Guida West and Rhoda Louis Blumberg, eds., *Women and Social Protest.* New York, 1990, p. 4.
3. Amy Schneidhorst, "'Little Old Ladies and Dangerous Women': Women's Peace and Social Justice Activism in Chicago, 1960–1975," *Peace & Change.* 26, 3, July 2001, p. 374.
4. Marian Mollin, "Communities of Resistance: Women and the Catholic Left of the Late 1960s," *Oral History Review.* 31, 2, 2004, p. 42.
5. For an interesting discussion of the link between masculinity and martialism, see William Ian Miller, *The Mystery of Courage.* Cambridge, MA, 2000.
6. Mollin, "Communities of Resistance," p. 50.
7. Barrie Thorne, "Women in the Draft Resistance Movement," *Sex Roles.* 1, 2, 1975, p. 184.
8. See Sara Evans groundbreaking study, *Personal Politics: The Roots of Women's Liberation in the Civil Rights Movement and the New Left.* New York, 1979.
9. Thorne, "Women in the Draft Resistance Movement," p.180.
10. Ruth Rosen, "The Feminist Revolution in California," in Marcia A. Eymann and Charles Wollenberg, eds., *What's Going On: California and the Vietnam Era.* Berkeley, CA, 2004, p. 86.
11. Harriet Gross, "Jane Kennedy: Making History through Moral Protest," *Frontiers: A Journal of Women's Studies* 2, 2, Summer 1977, p. 77.

12. Jo Freeman, "The Origins of the Women's Liberation Movement," *American Journal of Sociology* 78, 4, 1973, p. 800.

13. Nan Stone, quoted in Michael Foley, *Confronting the War Machine: Draft Resistance During the Vietnam War.* Chapel Hill, 2003, p. 185.

14. Francine du Plessix Gray noted that many in the Catholic Left characterized Philip Berrigan as a "desperado obsessed by the Gospel." Francine du Plessix Gray, *Divine Disobedience: Profiles in Catholic Radicalism.* New York, 1970, p. 78.

15. Mollin, "Communities of Resistance," first quotation p. 42, second quotation p. 44.

16. Campus Women for Peace, *Declaration of Opposition to the War and the Draft* [Berkeley, CA?], 1965. For information about the illegality of draft counseling, see "Counseling Draft Resistance: The Case for a Good Faith Belief Defense," *Yale Law Journal* 78, 6, May 1969, pp. 1008–1045.

17. Thorne, "Women in the Draft Resistance Movement," pp. 187–188.

18. Will Lissner, "6 War Protesters Seized on 5th Avenue," *New York Times,* 4 July 1969.

19. "Women Draft Disrupters Explain Action," *Guardian,* 6 September 1969.

20. Ibid.

21. Thorne, "Women in the Draft Resistance Movement," p. 190.

22. Foley, *Confronting the War Machine,* p. 182.

23. The slogan refers to men who refuse induction. Thorne, "Women in the Draft Resistance Movement," pp. 184, 190.

24. Mai Thi Chu, quoted in Campus Women for Peace, *For a Ceasefire in Vietnam* [Berkeley, CA?], 1965.

25. In fact, Mai Thi Chu's statement was quoted, but was not attributed to Chu, by Students for a Democratic Society. Jack Minnis, "Life with Lyndon in the Great Society," *Students for a Democratic Society Bulletin.* 4, 1, 1965, in *Online Archive of California.* http://content.cdlib.org/xtf/view?docId=kt5779n7hc&doc. view=frames&chunk.id= d0e305&toc.depth=1&toc.id=d0e302&brand=oac. Accessed 7 August 2007.

26. Valerie Clubb, interview by Dawn Walsh, 8 April 2000, in Missoula Women for Peace Oral History Project Collection (Missoula Women for Peace), box 1, folder 9, K. Ross Toole Archives, University of Montana, Missoula (UMM).

27. May McDonald, interview by Dawn Walsh, 27 February 2000, in Missoula Women for Peace, box 1, folder 3, UMM.

28. Eda Houwink statement, Record of the Women's Hearing on the War in Vietnam (Women's Hearing), 1972, p. 138, in Swarthmore College Peace Collection (SCPC).

29. May McDonald, quoted in Missoula Women for Peace Group Meeting, interview by Dawn Walsh, 28 February 2000, in Missoula Women for Peace, box 1, folder 14, UMM.

30. Anniece Allen statement, Women's Hearing, p. 101, SCPC.

31. Eda Houwink statement, Women's Hearing, 1972, p. 139, SCPC.

32. Amy Swerdlow, "'Not My Son, Not Your Son, Not Their Sons': Mothers Against the Vietnam Draft," in Melvin Small and William D. Hoover, eds., *Give Peace a Chance: Exploring the Vietnam Antiwar Movement*. Syracuse, NY, 1992, p. 160. For a more extensive analysis of Women Strike for Peace, see Amy Swerdlow, *Women Strike for Peace: Traditional Motherhood and Radical Politics in the 1960s*. Chicago, 1993.

33. Jeanne Molli, "Women's Peace Group Uses Feminine Tactics," *New York Times*. 19 April 1962.

34. Schneidhorst, "Little Old Ladies and Dangerous Women," p. 380.

35. The literature on nineteenth-century maternalist reform efforts is extensive. A good place to start is Theda Skocpol, *Protecting Soldiers and Mothers: The Political Origins of Social Policy in the United States*. Cambridge, MA, 1992, pp. 321–372. The literature on the ideology of separate spheres or female domesticity is also extensive. For an introduction to the ideology itself, see Jeanne Boydston, *Home and Work: Housework, Wages, and the Ideology of Labor in the Early Republic*. New York, 1990; Barbara Welter, "The Cult of True Womanhood: 1820–1860," *American Quarterly*. 18, 2, Summer 1966, pp. 151–174.

36. Florence Johnson statement, Women's Hearing, p. 149, SCPC.

37. Lois Barrett statement, Women's Hearing, p. 133, SCPC.

38. Sandra Perrin, quoted in Missoula Women for Peace Group Meeting, interview by Dawn Walsh, 28 February 2000, in Missoula Women for Peace, box 1, folder 14, UMM.

39. Sylvia D. Hoffert, *A History of Gender in America: Essays, Documents, and Articles*. Upper Saddle River, NJ, 2003, p. 210.

40. Elizabeth Ewen, *Immigrant Women in the Land of Dollars: Life and Culture on the Lower East Side, 1890–1925*. New York, 1985, p. 101.

41. Isabella C. M. Cunningham and Robert T. Green, "Purchasing Roles in the U.S. Family, 1955 and 1973," *Journal of Marketing*. 38, October 1974, pp. 61–81.

42. Another Mother for Peace mailer, n.d., in Social Movements Collection (SMC), box 6, folder 25, Vietnam Archive, Texas Tech University (TTU). Available online through Texas Tech University's *Virtual Vietnam Archive*. http://www.virtualarchive.vietnam.ttu.edu/virtualarchiv.

43. Letter from Jan Braslard[?], November 1972, in SMC, box 6, folder 25, TTU.

44. Letter from Dorothy B. Jones, [n.d.], in SMC, box 6, folder 25, TTU.

45. Alberta Slavin statement, Women's Hearing, p. 29, SCPC.

46. Mary Jane Badenoch statement, Women's Hearing, p. 16, SCPC.

47. Oral Lee Malone statement, Women's Hearing, p. 46, SCPC.

48. "Women and the Draft," n.d., in SMC, box 6, folder 31, TTU.

49. David Krieger, "The Another Mother for Peace Consumer Campaign— A Campaign that Failed," *Journal of Peace Research*. 8, 2, 1971, p. 163.

50. Jane Kennedy, quoted in Gross, "Jane Kennedy," pp. 77–78.

51. "Beaver 55 Strikes Again," in Alexander Bloom and Wini Breines, eds., *"Takin' it to the Streets": A Sixties Reader.* New York, 1995, p. 252; Gross, "Jane Kennedy," p. 73.
52. Donna Allen and Cora Weiss, quoted in Tom Wells, *The War Within: America's Battle over Vietnam.* Berkeley, CA, 1994, pp. 84–85.
53. L. B. Johnson, quoted in Doris Kearns Goodwin, *Lyndon Johnson and the American Dream.* New York, 1976, p. 278.
54. "Women and the Draft," n.d., in SMC, box 6, folder 31, TTU.
55. Catherine Stenger statement, Women's Hearing, p. 12, SCPC.
56. Gerald Gill, "From Maternal Pacifism to Revolutionary Solidarity: African-American Women's Opposition to the Vietnam War," in Barbara Tischler, ed., *Sights on the Sixties,* New Brunswick, NJ, 1992, p. 189.
57. Gill, "From Maternal Pacifism to Revolutionary Solidarity," p. 190.
58. Coretta Scott King, quoted in Gill, "From Maternal Pacifism to Revolutionary Solidarity," p. 180.
59. Tanya Luna Mount, quoted in Vicki L. Ruiz, *From Out of the Shadows: Mexican Women in Twentieth-Century America.* New York, 1998, p. 115.
60. Jacqueline Mack statement, Women's Hearing, p. 138, SCPC.
61. Janet Mezzack, "'Without Manners You Are Nothing': Lady Bird Johnson, Eartha Kitt, and the Women Doers' Luncheon of January 18, 1968" *Presidential Studies Quarterly.* 20, 4, 1990, p. 749.
62. Gill, "From Maternal Pacifism to Revolutionary Solidarity," p. 178.
63. Pamela Ashley, "Anti-War Pickets [letter to editor]," *Ebony.* February 1968, p. 18.
64. Rosalynne Hughes statement, Women's Hearing, p. 131, SCPC.
65. Gill, "From Maternal Pacifism to Revolutionary Solidarity," p. 193.
66. "A Negro Mother Tells Why U.S. Cannot Win in Asia," *Muhammad Speaks.* 10 February 1967.
67. Gill, "From Maternal Pacifism to Revolutionary Solidarity," p. 181.
68. Elizabeth "Betita" Martinez, "A View from New Mexico: Recollections of *Movimiento* Left," *Monthly Review.* 54, 3, July-August 2002. http://www.monthlyreview.org/0702martinez.htm. Accessed 3 November 2006. For more information on Chicanas and the war, see George Mariscal, "Chicano and Chicana Experiences of the War," in Marcia A. Eymann and Charles Wollenberg, eds., *What's Going on? California and the Vietnam Era.* Berkeley, CA, 2004.
69. "Chicanas meet Indo-Chinese," *El Grito del Norte.* 5 June 1971; Cover photographs, *El Grito del Norte,* 29 August 1970.
70. Kathleen Blee, ed., *No Middle Ground: Women and Radical Protest.* New York, 1998, p. 4.
71. Ann Q. Niederlander, "Findings," Women's Hearing, pp. 1–2, SCPC. For newspaper accounts of the hearing, see Charles Oswald, "Women at Hearing Here Roundly Condemn War," *St. Louis Globe Democrat.* 19 June 1972; "Women at War Forum Urge Consumer Boycott," *St. Louis Post Dispatch.* 18 June 1972.

72. Kathie Sarachild, "Taking in the Images: A Record in Graphics of the Vietnam Era Soil for feminism," *Vietnam Generation.* 1, 3-4, 1989, p. 241.

73. Although women now serve in combat environments, official Pentagon policy prohibits women from serving in the "infantry and other positions in which the primary mission is to physically engage the enemy." Department of Defense, Office of the Under Secretary of Defense, Personnel and Readiness, "Representation within Occupations," in *Population and Representation in the Armed Services: Fiscal Year 2001.* Washington, D.C., March 2003. http://www.defenselink.mil/prhome/poprep2001/chapter3/chapter3_6.htm. Accessed: 8 August 2007.

CHAPTER 8

Mixed Messages: Gender, Peace, and the Mainstream Media in Australia, 1983–1984

Suellen Murray

In the early 1980s, the Australian women's peace movement staged two major protests: in 1983, the Pine Gap Women's Peace Camp, held in central Australia, highlighted the presence of the United States Pine Gap military base near Alice Springs, and in 1984, the Sound Women's Peace Camp, held south of Perth, Western Australia, focused on visits by U.S. war ships and submarines to the Stirling Naval Base in Cockburn Sound. Both locations were remote from the major centers on the east coast of Australia: one in the central outback, the other on the west coast. While neither of these sites of the women's peace camps lent themselves to wide public visibility, Pine Gap and Cockburn Sound were key locations for protesting the presence of war in a period of supposed peace. At this time, while Australia itself was not at war, through the porting of U.S. warships in Australian waters and the placement of U.S. military bases in places like Pine Gap, Australia played an important strategic role in maintaining the possibility of war and supporting U.S. Cold War policies. Significantly, and at least partly, because Australia was not at war, the populations local to Pine Gap and Cockburn Sound were largely unsympathetic to the presence of women protesting about war.

As the historians Marilyn Lake and Joy Damousi have argued in their accounts of Australia at war and Australians' resistance to war, women have been "confirmed in their position of marginality," unlike men, who "whether as combatants in overseas wars or as anti-war activists . . . were the pre-eminent subjects of history."[1] Indeed, Australia's national history is built around (male) participation in war, with Australia's record in World War I heralded as "the birthplace of the nation."[2] Women's involvement as noncombatants, on the other hand, has typically been understood in terms of passivity, powerlessness, and weakness. The Australian women's peace camps of the 1980s (and, indeed, other expressions of women's peace activism across the twentieth century) challenged these perceptions and added an important gendered dimension to Australian understandings of war.

Given the importance of masculine involvement in war and warrior culture to Australia's national identity, it is perhaps unsurprising that women's peace activism has not been a strong area of Australian scholarship. Malcolm Saunders and Ralph Summy provide a historical overview of the Australian peace movement from the Boer War until the early 1980s and make mention of the women's peace movement contributions over that time but with little attention paid to the latter period.[3] Early twentieth-century women peace activists received some attention in the literature, including that by Joy Damousi, who has written on socialist women's opposition to war,[4] and others, such as Marilyn Lake and Judith Smart, who have documented the work of feminist peace organizations during World War I.[5] Consideration has also been given to Communist women's involvement in antimilitarism and to "banning the bomb" after World War II.[6] The Vietnam War was to be the next rallying point for women's peace activism, and Australian scholars, including Ann Curthoys, Barry Young, and Siobhan McHugh, have documented women's involvement in opposing conscription as well as the war itself.[7] Through their radicalization and subsequent involvement in women's liberation, this conflict, more than any other, has contributed to shifts in the understanding of gender in Australia.[8] Verity Burgmann and Brendan Cairns have discussed the wider nuclear disarmament movement of the 1980s of which the women's peace camps discussed here were a part.[9] Others have discussed aspects of the women's peace movement, including the nature of their activities and their goals,[10] but how their efforts contributed to wider understandings of gender and war have yet to be fully addressed.

To make sense of the antiwar activism undertaken at the Australian women's peace camps and to see how it has contributed to awareness of noncombatants' experiences of war, I turn to media reports and other press commentary. There is an abundance of this material available, at least partly because media exposure was promoted by the peace activists themselves. At both Pine Gap and Cockburn Sound, while linked into their own communication networks, the peace activists relied primarily on the mainstream media to take their antiwar and antiviolence messages to the widest possible audiences.[11] In contrast, the means provided by electronic media that are accessible to contemporary activists, which allow for much more personal intervention and provide a possibility to circumvent the mainstream media, were not available at the time. Clearly, many of the protest actions at Pine Gap and Cockburn Sound were undertaken as "media events" to attract the media's interest and, in doing so, to draw attention to the serious issues that the women were addressing.

The relationship between the women and the media seemed, at the time, largely characterized by, at best, ambivalence and, more often, antagonism, reflecting to some extent particular views about women in Australian society. This was not totally unexpected because the women activists at Pine Gap and Cockburn Sound were aware that their predecessors had experienced similar ridicule, abuse, and attacks from those who did not agree with their points of view.[12] However, in this chapter, I argue that while many of the portrayals marginalized the peace activists and trivialized their activities, there were also several reports that took the women seriously and pointed to the strength and subversive power that they were wielding through their protests. In effect, I suggest that the "mixed messages" presented about the peace activists were part of a wider shift in the ways in which women were being understood in Australian society in the 1980s. Hence, I reveal not only the nature of the women's peace activism, their position as noncombatants, but also how media portrayals of them and their activities reflected and informed wider Australian understandings of gender.

The Australian Women's Peace Movement of the Mid-1980s

While the Australian women's peace movement has a long history, during the mid-1980s, there was a heightened level of activity.[13] At this time, a string of antiwar and antinuclear actions were undertaken worldwide, and the Australian women's peace camps were part of this international movement. Although Australia was "at peace," the focus of antiwar attention at this time was the Cold War and the threat of nuclear war. As noted by the historian Ann Curthoys, "the [Australian] peace movement swelled enormously in the 1980s as the very real possibility of nuclear war, and huge demonstrations, involving 80,000 people or more, were held in most major cities and towns."[14] Such demonstrations were attended by men as well as by women and reflected a society-wide recognition of the dangers of nuclear war, both for Australians and for the world at large.

The women's peace camps at Pine Gap and Cockburn Sound each lasted a fortnight and took place in November 1983 and December 1984, respectively. In between there was another women's peace camp at Salisbury Weapons Research Establishment near Adelaide, and a mixed (gender) camp was held near Roxby Downs uranium mine in South Australia. In 1985, North West Cape, the site of a U.S. military base in northern Western Australia, was the site of another mixed peace camp. Other major events were the annual Palm Sunday rallies held in all major Australian cities by People for Nuclear Disarmament.[15] By the mid-1980s, these rallies were the largest protests in Australian history. In 1985, 350,000 people marched in support of peace across the country.[16] Another regular activity of the Australian peace movement was the marking, by means of rallies and marches, of the anniversaries of the bombings of the Japanese cities Hiroshima and Nagasaki in 1945.

In addition to these major events, there was a stream of smaller actions drawing attention to local peace and military issues. For example, in Western Australia, which at this time received the majority of U.S. warship visits to Australia, peace activists protested about their presence and the associated risks that they created to the local populations and also by involving Australia in the wider international conflict between the United States and the Soviet Union. There were local protests that supported the larger actions as well, including, for example, camps set up in Canberra and Melbourne and other actions in Perth, Hobart, and other towns and cities around Australia in support of the Pine Gap and Cockburn Sound women's peace camps.[17]

In several respects, therefore, the women's peace camps at Pine Gap and Cockburn Sound were not extraordinary. They occurred within the context of a much larger and popular nationwide phenomenon that was campaigning against the Cold War and the capacity of the United States and the Soviet Union to initiate a global nuclear war. Even though it was geographically removed from these nations, Australia was involved in the proliferation of the Cold War and nuclear threat in a number of ways. First, Australia was (and still is) a partner in the ANZUS (Australia New Zealand United States) defense treaty, which enabled the siting of U.S. military bases in Australia and the visits of U.S. warships and submarines to Australian ports.[18] While never confirmed by the United States, it was firmly believed by those in the peace movement that many of these vessels were nuclear powered and that some were nuclear armed. The peace movement activists were concerned that by allowing the visits, Australia actively supported U.S. militarism and indirectly caused itself to become a target for potential nuclear attack by the Soviet Union. Another fear was the potential for nuclear accidents putting the Australian population and environment at risk. Moreover, Australia mined and exported uranium, a necessary ingredient for nuclear weapons. In terms of how the peace protesters saw it, while this trade was lucrative, it also endangered the safety of the world. Nuclear war obviously did not differentiate between combatants and noncombatants: it would affect everyone, not just those in battle.

It was all these concerns about Australia's involvement in nuclear war that led women to join the women's peace movement and to participate in its two key events of the 1980s—the women's peace camps at Pine Gap and Cockburn Sound. However, many women who joined the Australian women's peace movement in the 1980s were not involved in the wider antinuclear peace movement but came because of their links to the Australian women's movement or other social and environmental protest actions, such as those opposing the logging of native forests. Others joined from their involvement in left-wing politics. Many were inspired by women's peace camps elsewhere, in particular, at Greenham Common in England, which commenced in 1981 in protest at the installation of cruise nuclear missiles there and continued until 2000.[19] Therefore, while we can situate the Pine Gap and Cockburn Sound peace camps within the wider Australian antiwar campaign of the time, it is important to see these women's camps as unique as well.

While not mutually exclusive or uncontested, Australian women's motivations for setting up and participating in the peace camps can be located within two frameworks of understandings. First, an interest in peace was maternalist in origin and related to a belief that women were nurturing and protective. Second, an interest in peace and the nuclear disarmament movement was framed within particular forms of feminism and, in particular, radical feminism and ecofeminism, in which war was understood as a form of patriarchal oppression.[20] As Curthoys states, "feminists often portrayed militarism and ecological waste and destruction as the product of male power, as well as masculine values and priorities."[21] A significant difference, then, between the general peace movement and the women's peace movement was the latter's emphasis on wider issues of violence and, especially, violence against women. Thus, the women's peace movement was not just concerned with war and warfare but also with what was understood as "war against women," including domestic violence, sexual assault, and the impact of economic and social disadvantage.

The diversity in motivations for joining and organizing the peace camps at Pine Gap and Cockburn Sound is clearly reflected in the number of groups that participated. The peace camp at Pine Gap, for example, was jointly organized by the Women's Action Against Global Violence based in Sydney and the local Alice Springs' women's peace group, alongside input from the other state-based groups.[22] In Western Australia, Women's Action for Nuclear Disarmament (WAND), based in Perth, organized the Cockburn Sound camp, with support from women's groups in other states, particularly in terms of the logistics of transporting women from all over the country to the west coast, which is more than 3,000 kilometers away from Australia's main cities on the east coast. Over the course of the mid-1980s, these various women's peace groups changed their name to Women for Survival, in line with the national trend for a uniform name for Australian women's peace and antiviolence groups and also reflecting interests wider than nuclear disarmament.

While on the surface it may seem counterproductive to organize protests aimed at gaining maximum media exposure in remote locations, the women at Pine Gap and Cockburn Sound had specific reasons for situating the camps where they did. In both cases, the presence of nearby military installations and naval bases enabled the camps to easily highlight and focus their protest activities on Australia's engagement in the Cold War. As a result, the "secretive" Pine Gap satellite telecommunications military base ("probably the most important outside the US" and "run by the CIA")[23] was an ideal choice as was Stirling Naval Base in Cockburn Sound and Perth's port city of Fremantle, which witnessed the porting of nuclear-powered and nuclear-armed U.S. warships. The proximity of these military targets enabled the peace activists to take their protests directly to the source of their concerns. However, the women aimed to do more than just let Australians know of the existence of Pine Gap and the warship visits. Women for Survival were also opposed to the existence of all foreign military bases in Australia and sought to end foreign intervention in Australian affairs. By the time of the Cockburn Sound peace camp, this goal was explicitly stated as "an end to the US-Australian military alliance (the ANZUS treaty)."[24] The women were concerned about violations of the United Nations declared Indian Ocean Zone of Peace and aimed to draw attention to the militarization of the Indian and Pacific Oceans. At Pine Gap there was stated support for the struggle in Europe against the siting of cruise and Pershing missiles, in solidarity with the women at Greenham Common particularly. During the course of the Pine Gap camp, these missiles were installed at Greenham Common. Women for Survival also opposed Australia's involvement in any stage of the nuclear cycle, including uranium mining, nuclear power generation, or nuclear weapons manufacture.

However, the goals of the camps were not only concerned with antiwar and antinuclear themes. The women at the peace camps came from a wider antiviolence political background, and their aims also included the support of Aboriginal land rights and the redirection of defense spending into areas of social need and the protection of the environment. They also supported "women and children of all races and cultures in their struggle against violence and oppression," and, at Cockburn Sound specifically, there were concerns about the impact on the local communities of the 8,000 sailors that arrived during U.S. ship visits. Hence, Women for Survival protested against "the exploitation of women resulting from the use of Cockburn Sound and Fremantle as rest and recreation ports for US military personnel."[25] The presence of military personnel was another way in which war impacted on a peacetime society, but whereas during World War II when there had been largely romantic ideas about Australian women and visiting U.S. sailors, during the 1980s the dangers of fraternization were emphasized by the women's peace camps. In contrast, local residents were keen to get to know the visiting sailors and resisted challenges by the women activists.[26]

Working with the Media

The Australian women's peace camps did not exist in isolation. They were supported by the women's peace organizations in the different states and territories and by their much wider networks of supporters.[27] As noted already, it was from this base that other local actions were launched both at the time of the camps and in the intervening and following years. Importantly, however, it was through the state-based women's peace groups that the goals of the movement were promoted and its activities advertised. They published newsletters and sent them out widely, specifically targeting both individuals and other women's organizations. Information about their aims and activities was also provided to like-minded publications as well as to the mainstream press.[28] Evidence of the effectiveness of these activities is found among the many letters and telegrams of support received at the peace camps.[29] There were also supportive letters to the editor published in the Australian press, although, many more were unsympathetic.

Other creative ways in which their ideas and interests were circulated were through postcards, T-shirts, fliers, banners, brochures, and posters.[30] After Pine Gap, a booklet of photographs was produced, a photographic exhibition was organized, and a film was made at the Cockburn Sound peace camp to raise awareness of the protest activities. Public meetings were held in the lead-up to the camps.[31] Support also came from prominent politicians, who were willing to speak in support of the women's peace movement, including in Parliament. For example, Kay Hallahan, Labor member of the Western Australian Legislative Council, provided assistance to WAND in the lead-up to the Cockburn Sound peace camp when P. H. Wells, a Liberal member, claimed that WAND was discriminatory in their activities by not providing information to male members of the parliament. In her speech to the parliament, Hallahan minimized these concerns and noted that WAND was a group of women with a real concern about nuclear disarmament . . . putting out very good information and particularly drawing young women together in order to explore how we can move towards nuclear disarmament.[32]

Clearly, then, the women peace activists were reaching out to networks beyond their traditional supporters, and they were attempting to inform the wider Australian community about their concerns about war and violence.

While their own networks and other allies attempted to take their message to a wider audience, the women's peace activists were also aware of the need to engage with the mainstream media.[33] At both peace camps (and within the state-based women's peace organizations) media collectives were established to promote and manage relationships with the media.[34] As described in the information kit provided to all participants at the Cockburn Sound camp, the media collective was "a buffer between the media and the camp" and (female) media representatives were to be allowed in to the camp only on certain days and only under escort. The media collective was responsible for maintaining communication with members of the media, including writing press releases, providing other information about the aims and activities of the camps, and nurturing relationships with sympathetic journalists. Consistent with the philosophy of collectives and in opposition to what was considered to be masculinist and hierarchically oriented political processes, consensus decision making was used and tasks were shared. Hence, the media collective was to be "responsive to the needs and feelings of the whole camp," and spokeswomen were to change daily. Individuals could speak for themselves, but they were expected to make it clear that they were not speaking for the peace camp as a whole.[35] These processes were not without their difficulties, partly because of the large numbers of women involved (there were up to 1,000 women at each of the camps, although the media collectives were much smaller groups of around a dozen women), the time it took to discuss and come to decisions, and the varied knowledge and skills held by the activists, all issues not uncommon to other collectives.[36]

The most controversial aspect of the relationship between the media and the peace activists was the latter's decision to allow only female journalists into the campsites. This policy was in line with the "women-only nature of the camp" and as "a form of affirmative action for women in the media."[37] The policy, however, did not mean that the activists would not engage with male journalists at all, rather that they did not want them to come into the campsite space. As I will discuss shortly, there was little sympathy from the press for the idea that women should gather in public spaces and, at the same time, live privately within these spaces without what they considered to be the constraints of men's presence. Some in the media took the request that men not enter the private living space of the campsites as an affront to their freedom of movement and as an insult to their masculinity. Bruce Stannard, who reported on the Pine Gap camp, complained that "while female journalists were literally welcomed with open arms, the women [peace activists] demanded that the 20-odd men covering the event for television, radio and newspapers throughout Australia carried special accreditation documents."[38] Norm Taylor, president of the Western Australian branch of the Australian Journalists' Association, in response to WAND's request that media outlets send female crews and for male journalists not to enter the Cockburn Sound campsite, stated that they "couldn't support a situation where any group is discriminating against a section of our membership simply because they happen to be male."[39] That these men should think of themselves as being "discriminated" against suggests a lack of insight into the nature of discrimination and the ways that female journalists, at this time, were very likely to have experienced exclusionary practices as part of their day-to-day employment (Equal Opportunity legislation was only just being introduced in most Australian jurisdictions during the mid-1980s). Nonetheless, some female journalists also expressed concerns about denying access to male journalists.[40] While the peace activists had not expected an overly positive relationship with the media, among some press representatives the women-only media policy provided the ammunition for them to express outright hostility toward the peace camps, a development that did not assist in getting the activists' message across to the wider public.[41]

The women-only nature of the camp did not simply raise the interest (and ire) of journalists. Because it was women-only, it was assumed by some that many of the camp's participants, if not all, were lesbians, and that somehow lesbians were innately aggressive, fearful, and loathsome. As reported in the *Sound Telegraph,* a community newspaper local to the Cockburn Sound peace camp, and reflected in other papers local to both the Cockburn Sound and Pine Gap peace camps:

> With a great deal of trepidation this cowardly reporter donned her best gardening gear for an inside look at the Sound Peace Camp . . . So, with quaking heart, and visions of the lesbian element keeping guard, I pushed my way through the bushes for "come what may" . . . On one side is the strong lesbian element, butch hairstyles and boilersuits and on the other those who genuinely wanted a peaceful protest.[42]

Instead of emphasizing the toughness of the women (characterized by butch lesbianism), an alternative view of the women-only nature of the peace camps focused on their helplessness. This approach provided ample opportunity for cartoonists to ridicule the camp participants with images playing on the women's implied powerlessness and weakness. In one cartoon published at the time of the Cockburn Sound camp, two women peace activists are depicted running away from a mouse while a burly man looks on. As they run, one woman asks the other: "Do you think it would compromise our position if we asked him to defend us?"[43] This stereotyped form of femininity, requiring the aid of men to defend women's safety and well-being, was obviously being challenged by the mere existence of a women's peace camp. Here were groups of women protecting themselves (with the "lesbian guard"!) and also attempting to promote the protection of others not through means of war and violence but through peace. The camps further challenged ideas about women's location in society by purposefully situating themselves in public space in contrast to the typically feminine (and private) domain of the home. These women, instead of being "on the home front," were occupying the "battle" ground, a space traditionally inhabited by men as warriors. As we shall see, space was to be a further issue because it was not just that the women inhabited public space, but they also forcibly invaded restricted military space.

A further way that the women challenged stereotypical ideas of femininity concerned their appearance. While the reporter quoted above was not alone in referring to the hairstyles and clothes of some of the women ("butch hairstyles and boilersuits"), others also made references to their lack of cleanliness and the unhygienic conditions under which they were living, conditions deemed to be most unladylike. For example, the Cockburn Sound camp was described as "a grubby little event," and it was suggested that its participants should at least look "as if they bathed once in a while."[44] Despite the difficult living conditions, the media expected that the women, principally because they were women, should maintain acceptable levels of feminine hygiene and cleanliness.

By challenging stereotypes about the ways that women should live their lives, the peace activists created what Barbara Brook has called "enormous crises of interpretation," and these were further evident in other portrayals of the women.[45] The media gave considerable attention to the perceived reasons for women's involvement in peace activism, often in polemical terms. While less stereotypically feminine women were highlighted from very early on, there was also a focus on women who were deemed to be promoting more maternalist goals. Some journalists glorified individual women activists by making reference to their maternal status, including a woman who was identified as a "grandmother" and another who was described as breastfeeding at the time of her interview. In stark contrast, the media also made much of women, particularly "aggressive young women," who were more "like screaming savages" and had, at least in the eyes of the reporters involved, lost all their "natural" female, let alone maternal, sensitivities.[46] Such references purposely demeaned the women but were not unique to Australian representations of women's peace movements. Similar patterns of representation are found in the coverage of Greenham Common by the British and international press and highlight the ways in which war, even during peacetime, reifies some forms of gendered representations and attitudes toward women.[47] While the disparaging comments may reflect media (and wider public) distaste for the women's peace movement and their activities, they are also an indication of more deeply rooted views about women and their role in Australian society and, in particular, about those feminist women who were challenging traditional views of femininity and their relationship to war. It was precisely such traditional views—that women could, or rather should, only be maternal and protective in their roles as peace-seekers—that the peace activists sought to challenge and overturn.

However, even when journalists depicted the women activists as "maternalist" peace activists ("those who genuinely wanted a peaceful protest"), they typically did not take them seriously. The press seemed surprised to discover that they were not just going to sit around discussing the nuclear risks to their children and grandchildren for the duration of the camps. They (like the "screaming savages" and the "aggressive young women") were also there to take action, and those actions included making forceful statements about the risk of nuclear war and the wider social environment in which it took place. Moreover, these statements were accompanied by activities that were considered "unfeminine" and, at times, illegal or, worse still, contradictory: how could these women profess to be peace activists when they used "violence" to seek peace? Not surprisingly, what was considered "violent" was another source of tension between the peace activists and the press, and indeed, among the wider public.[48]

In terms of how the Australian media presented them, the greatest condemnations were reserved for the activities of the women that violated the military bases, and particularly their fences, near the peace camps. Margaret Laware, in her analysis of protest actions at Greenham Common, notes the use of strategies of subversion to destabilize the fence surrounding the military base there and, in doing so, to "undermine its symbolic significance as a marker of military space."[49] The most obvious subversive use of the barrier was to remove it by cutting it and going through it or by climbing over it, in acts that Laware has named "reclamations." She argues that these strategies appropriated the fence and impacted on the authority and power of the military base.[50] Through "violating" such fences, the women activists, both at Greenham Common and at the Australian peace camps, highlighted the existence of the military bases (and the functions of the enclosed military space) and, perhaps more importantly, disrupted (if not corrupted) the significance of these places and spaces. The peace activists decorated the fences with ribbons, banners and placards and graffitied the adjoining road with peace and antiviolence slogans, appropriating them for their antiwar purposes. They made explicit reference to their reasons for doing so: "We women at Cockburn focus on the gate as symbolic of militaristic barriers. We accept our responsibility to act against the injustice, the immorality, the affront to humanity represented by these barriers."[51] Moreover, the presence of hundreds of women conducting their daily lives as well as protesting by singing, engaging in street theater, and dancing within close proximity offered a visual tribute to the "reclamation."

The press, however, did not see such actions as "reclamations." Instead, typically, they saw them as acts of violence, which resulted in them questioning the legitimacy of the women's protest. Over the course of the peace camps, sympathy in the media for the women's positions declined. Alternatively, having identified that some of the women were peace-loving mothers and grandmothers, who would not compromise their pacifist principles by participating in "violent" protests, when "violent" protests did occur, some journalists (erroneously) concluded that that there must be divisions in the camps. This led to newspaper reports suggesting that women were leaving the camps in large numbers because they were dissatisfied with the choices made by their activist colleagues.[52]

The media further questioned the peace activists' motivations and undermined their validity by alluding to the possibility that the women were Soviet stooges. For example, Michael Barnard in the *Northern Territory News* stated that Women for Survival had "labyrinthine links" with "hard-Left groups and individuals" including "a number of communist-led unions" and "outright Soviet-policy apologists such as the Australian Peace Committee."[53] In the *Centralian Advocate,* Bernie Kilgariff, Northern Territory Senator, in discussing the Pine Gap camp, was reported as suggesting that the funding for the "so-called peace movement" was linked to Soviet sources.[54] While most reporting did not overtly make such links, letters to the editor made these claims, especially in the newspapers local to the two peace camps in the Northern Territory and Western Australia. For example, a "grandmother of five and great grandmother of one," a resident of Safety Bay, a suburb adjacent to the site of the Cockburn Sound camp, wrote to the *Daily News* that "she was born in England in the Great War, married and had my children in the World War, and I'd rather die in the next war than live under communist rule."[55] Another letter-writer, "a peaceful mother of three," a resident of a northern suburb of Perth, asked: "Does the women's movement have a date yet for their camp in Moscow?"[56] In effect, the implications were that the women were untrustworthy or, at the very least, naive.[57] Worse still, they could be agents of the "enemy" in the Cold War, reflecting wider attitudes that the women were mindless and could be duped by others. In making these links between the peace activists and the USSR, the women were treated as enemies that could be complicit in initiating a war led by the USSR, and thus, the peace activists were placed in the position of being at war against their fellow Australians. This obviously turned the real reason behind the peace camps on its head and justified all manner of condemnation.

At both Cockburn Sound and Pine Gap, the local papers were the most virulent in their criticism of the peace activists, which was strongly reflected in the letters to the editor written by members of the local community. This is perhaps not surprising given that the local economies relied heavily on the continuation of those very things about which the peace activists were protesting. The Pine Gap base provided employment to Alice Springs' residents and contributed to its economy through the supplying of the base and its staff. The U.S. Navy was calculated to contribute in the order of $14 million per annum to the Western Australian economy at the time.[58] Moreover, both camps were held in small, isolated communities that were unfamiliar with peace activism or the activities of the women's movement, and this, too, could have added to local distaste for the camps.

Of course, while all such negative press was disappointing from the point of view of the women at the peace camps, the "chief paradox" of their situation was, as noted by Julie Emberley and Donna Landrey in relation to Greenham Common, that "the media are crucial to the camp's effectiveness in consciousness raising on a mass scale yet remain its worst enemies."[59] Without press coverage, the peace activists' messages could not circulate and the camps would have been a failure. To this end, the peace camps staged numerous events to take advantage of the media attention they were receiving. While there may have been a generalized view among the peace activists that "the media are far from neutral bystanders . . . they may not work directly for the ruling class [but] they certainly do not work for social movements,"[60] it did not prevent them attempting to work with the media to get their messages across. Indeed, there may have been much greater success in doing this than realized at the time.

Taking Action: "Closing the Gap" and
"Breaking the Sound Barrier"

Participants writing shortly after the two camps were over remembered the media coverage as having been "hostile."[61] I was a participant at both of the women's peace camps, and, had I been asked at the time, I would also have thought similarly, although I knew of at least one journalist who was a close friend (and ally) of other peace campers, who worked hard to get our points of view across in the daily tabloid for which she worked. In reviewing the press coverage from the vantage point of more than two decades later, however, I see much greater diversity and complexity in the responses of the media to the activities of the peace activists, reflecting possible shifts in understandings of women and their relationship to war. These media representations, while still depicting traditional understandings of gender, also offered key markers of change.

The first major action organized at the Pine Gap campsite occurred on 13 November 1983, the third day of the camp, and resulted in the arrest of 111 women. Most of the women gave their name as "Karen Silkwood" to the police, in memory of the American antinuclear campaigner who had worked at a U.S. nuclear power station and who died in suspicious circumstances in 1974. In addition to the explanatory comments regarding the name Karen Silkwood, some newspapers also reported on the issues raised by the women over the course of their action. For example, it was reported that one of the Karen Silkwoods, from within the perimeter fence of Pine Gap, stated that the base was "a daily example of Australia's recolonization by a foreign power."[62] By speaking from the restricted space of the military base—a space to which most Australians were denied access—she highlighted the impact of war on peacetime Australia.

The novelty of the names of the arrested protestors attracted much media attention and overrode some concerns that the women had trespassed and vandalized U.S. military property.[63] It was also noted that the police has shown considerable reluctance to arrest the women, and they reportedly stated that "it was all pretty genteel . . . there was no violence, no hassles, no worries."[64] The Alice Springs tabloid *Centralian Advocate* called it a "gentle ambush."[65] However, the next major action two days later, named "Closing the Gap," witnessed harsher treatment of the arrestees by the Northern Territory police, and this also captured the media's attention.[66] An entry from my journal written at the time describes what happened:

> We gathered at the gates singing "Take the toys from the boys". Professor Wipemout arrived with the nuclear missile on the Mini-moke with a

camouflage-clad guard. Speeches, singing, clapping, jeering and the missile was brought up. Lauri spoke about cruise missiles being deployed at Greenham today. The missile was disarmed: balloons released, nose dented, then carried forward, all the time we were singing "Take it back, it's a load of crap, turn it into scrap, take it back". Women attempting to climb over the fences with the missile to take it back to those in command at Pine Gap but they were pushed back, thrown back, over the fence, arms and legs flailing, by the line of police lining the gate. Some women climbed through the fence and ran forward up the road with the crumpled missile to be arrested by the waiting group of police.

At the same time, women pushing on the gate, five deep. Police lining the gate pushing back arms outstretched . . . The right gate came off its hinges, a moment's hesitation and then turning to the left hand gate, and twisting and pushing it off its hinges. Several women had run forward to be blocked by the line of police who then formed a human fence. Women took the gate away, quickly painted a banner, "Preventative medicine: doctors for peace", and a woman was ceremoniously brought back on the gate being used as a stretcher . . . Women slowly began to form a line facing the line of police, our arms linked singing "We are gentle, angry women and we are singing for our lives" . . . The gate that had been removed was soon replaced by the police to be quickly daubed with graffiti stating "Congratulations boys on a hasty erection".[67]

The press reporting on "Closing the Gap" focused on the confrontation between police and protesters, the removal of the gate, and the speed with which the protestors were arrested. Importantly, the women's statement concerning the deployment of cruise missiles was successfully (although less prominently) noted in the press. For example, Sydney's daily tabloid the *Daily Telegraph* reported on its front page that

two hundred women broke into the top secret Pine Gap military base yesterday after ripping a metal gate off its hinges and smashing through a wall of police. The violence erupted after the women staged an early morning anti-nuclear demonstration to coincide with the installation of cruise missiles at Greenham Common in England.[68]

Sydney's daily broadsheet the *Sydney Morning Herald* also covered the protest with a front-page photograph, captioned "We are strong—we say No to the Bomb . . . About 100 women pulled the gates down in protest against the deployment of missiles in Europe."[69] The determination of the women is reflected in other photographs in newspapers across the country that day, and while they could be read simply as acts of violence by a group of aggressive women, they also suggest something about the activists' physical strength and moral commitment to their cause. In doing so, the press, perhaps inadvertently, helped to challenge traditional female stereotypes of weakness and mindlessness.[70]

In addition to reports on the Australian protests, Australian newspapers also covered protests in the UK about the deployment of the cruise missiles.[71] Here again, the coverage focused as much on the "violent" actions taken by British female activists as it acknowledged the point that was being made about the deployment of cruise missiles. In both cases, it reflects a willingness on behalf of the newspapers to take the women and their messages seriously. Wider support from the Australian public was also reflected in letters to the editor. For example, Sylvia Monk from Queensland wrote that

> we should all be saying "no" to the lies that we can save ourselves by preparing to kill others and ourselves, whether these lies are told in Paris, Moscow, Washington, Pretoria or Alice Springs, and we should not be leaving a protest for peace to these women at Pine Gap.[72]

A concern with media reporting of the issues about which the peace activists were protesting was highlighted in a cartoon in Melbourne's daily broadsheet the *Age,* later that week.[73] In this cartoon, a stereotypical lesbian feminist (with spiky hair and overalls) at Pine Gap is yelling at a barrage of microphones and camera lenses, "Why don't the media report the real issue?" She then goes on to outline the mass of concerns including "the way that the male-dominated nuclear multinational chauvinist complex use militarist police brutality and media oligarchies to competitively rape sexual liberties and Aboriginal identity." At one and the same time, the cartoon satirizes the peace activists *and* summarizes their frustrations with the press. Even though the cartoon used a stereotyped image to categorize the women activists ("butch hairstyles and boilersuits"), it also successfully conveys something of their antiwar and antiviolence messages as well as of their strength and determination.

In much the same way as Australian newspapers presented the "Closing the Gap" campaign at Pine Gap camp in ambivalent terms, they would also send mixed messages about Cockburn Sound's major protest actions. On the third day of the Cockburn Sound camp on 6 December 1984, several hundred singing, chanting, and dancing women marched to the gates of the Stirling Naval Base. Once there, the women presented speeches and then successfully sought a meeting with the naval commander. At this meeting, they requested to meet with the minister for defense, and the naval commander agreed to pass on their request and concerns about Australia's peacetime participation in the support of war. The women then formed a vigil at the gate to await the arrival of the minister. Earlier that day, street theater performed by the women satirized the U.S. Navy. The reporting of this first major action, while titled "Vandals spoil protest gains" in the *West Australian* and including some discussion of the graffiti that was painted on the road to the gates of naval base, also noted that "the Point Peron peace camp demonstrators achieved a major victory." In particular, it listed in full the demands presented to the government by the activists.[74] This action received coverage in other Australian press, and many of their demands were noted.[75]

It would seem, then, that the women were being taken seriously, and their antiwar and antiviolence messages were being heard (if not understood). However, the next major action, "Breaking the Sound barrier," occurred three days later but did not receive the relatively positive media attention that this first action had received. An entry from my journal at the time of the Cockburn Sound peace camp describes this event:

> The procession left through the camp with singing, music, banners, face paint, streamers, inflatable rubber ducks, liloes and surf boards to the gates of the naval base . . . We sang, Biff spoke about male violence, boys' toys and militarism. There was street theatre to similar effect, a black ball as a bomb, a painted blue and green ball as the world and a maze game with questions as cues. At the same time, some women set off to swim to the sentry box past the surprised line of police. The "Bounty" [a boat owned and crewed by peace campers] cruised up and down and around the causeway giving the water police a merry chase and generally keeping an eye on the women in the water. At the [two-meter-high] gate, women were clambering over with the assistance of milk crates and other women's outstretched supporting arms, at the same time as being pushed back by police. Women were being held bodily in the air at the height of the gate as they tried to make their way over . . . After getting over the gate, they ran up the road and were chased, arm-locked and arrested by the police.
>
> That night we slept on the lawn outside Fremantle police station maintaining a vigil while women were held inside. In the morning we found the women around the back and we spoke with them through the walls and windows, waiting for them to go to court. We had changed the *West Australian*'s headline sheets of "75 arrests in women's assault on island" to "75 arrests in women's celebrations" and "75 arrests in women-won island."[76]

The "Breaking the Sound barrier" action was widely reported in the Western Australian press and focused on the "frenzied battle" and the "demonstrators' assault" that "resembled a game of human volleyball."[77] While later reporting focused on the "violence" of these events and community disquiet about the protestors' actions, in the earliest reporting there was some acknowledgement of the reasons for the actions.[78] However, little attention was paid to the specific issue of U.S. ship visits and their associated nuclear risks or of the impact of sailors on local communities. Instead, the action was criticized as violent in contradiction to the nonviolent intentions of the camp and against the wishes of many of the camp's members.[79] The staging effects were also disparaged: "[the action] smacked of a Hollywood production. Every move was well orchestrated."[80] The local press noted that the actions themselves were not the only opportunity that the women had to make statements about their antiwar feelings, but such expression of these views was treated disdainfully. For example, in a "rowdy" session in Fremantle Magistrates' Court, "many of the women tried to read anti-nuclear statements to the court but were silenced by the magistrate . . . In the afternoon there was a brief scuffle between two policemen and a woman who was reading a statement."[81]

While the women were depicted in these reports as wild and unruly, their behavior can also be read as actions of those who were highly committed and willing to speak out about what they considered to be unjust and wrong. Some publicly acknowledged the peace activists' commitment and sincerity, as indicated in a letter to the editor of the *West Australian*, written by W. Hartley of Perth:

> That these women have made the long trek from their homes and accepted the discomforts of their present situation leaves little doubt in the minds of reasonable people that their motives are sincere and commendable and call for the admiration of those unblended by prejudice.[82]

Others also challenged the negative perceptions presented in some of the press and seemed to understand something of what the women were trying to achieve. For example, Ailsa Ruse of Perth commented in a letter to the editor:

> As a conventional old lady, I cannot fail to admit that the appearance and behaviour of the demonstrators near the Stirling naval base "get up my nose" . . . However, as a woman, mother and grandmother, I quietly applaud them. They are worthy successors to suffragettes, without whose antics women would never have had a vote.[83]

In line with these more sympathetic views about the peace camp, at least one interstate newspaper successfully presented the peace activists' antiwar and antiviolence messages and, in doing so, portrayed the women themselves as thoughtful and determined. The *Age* described the peace camp as "non-violent throughout the confrontation" and quoted a camp spokes-woman as saying that the "the fence symbolises the division created between people by nuclear militarism which claims to protect us while in fact expos-ing us to violence." Furthermore, a reason for the action was "to link with women all over the world who want to reclaim the earth's resources to affirm life rather than destroy it."[84] These varying perceptions, both in terms of whether the action was viewed as violent or not in terms of the inclusion or exclusion of commentary about the intended meaning of the actions, are characteristic of the "mixed messages" about gender, war, and peace that were offered in the press coverage of both the women's peace camps.

Mixed Messages

The media coverage of the peace camps at Cockburn Sound and Pine Gap provides an opportunity to consider the ways in which gender was under-stood in relationship to war and peace during the mid-1980s in Australia, particularly in relation to noncombatants and their resistance to war and violence. The peace camps did attract media interest, and media responses to the women's peace activism varied considerably: some were, as expected by the women, hostile, and this hostility emerged in two ways. Hostility could arise, first, because the women were deemed unfeminine and that they were challenging gender expectations, for example, by excluding men, by being "dirty," or by questioning their (hetero)sexuality, or, second, because their antiwar messages were clouded by "violent" protest action. In retrospect, perhaps, we should not be surprised that, at times, the press focused on the women's actions (and their appearance) rather than on their messages. The protest activities were playful—sometimes zany—and often intentionally humorous. Is it a wonder that the journalists (and their edi-tors) sometimes got lost in the theater, viewing it as "violent" and "unfemi-nine" rather than as a way of reclaiming the space, as Margaret Laware would suggest, and, consequently, missed the message? Is it surprising that the mainstream press was skeptical about the activities of a thousand women protesting about both nuclear and patriarchal threats?

Indeed, what is now surprising is the attention that was given to what they were trying to achieve—the dissemination of messages about the threats of both nuclear war and the impact of aggressive forms of masculin-ity, especially in the interstate papers, that suggests a willingness to look beyond the stagecraft and the somewhat threatening messages. So, my analysis suggests a more complex story than uniformly antagonistic responses to the protests at Pine Gap and Cockburn Sound. While not consistent or comprehensive in their approach, many newspaper reports did include some commentary on the intentions of the peace activists and their concerns about violence and war. Moreover, they portrayed them as thoughtful, committed, strong, active, and determined, in contrast to more typically feminine characterizations of passivity, mindlessness, and weakness.

The Pine Gap and Cockburn Sound camps never developed the fame that Greenham Common did, partly, of course, because of their lack of relative longevity. Greenham came to mean both "a muddy encampment of antinuke women sixty miles from London in the Berkshire countryside" as well as "a continuing protest against the deployment of US military hardware in Britain and Europe, and, more metaphorically, against the masculine economy of aggression, militarism and global violence that nuclear weapons metonymize."[85] The Australian women's peace camps were also framed within this latter, wider meaning—that it was not just the nuclear weapons about which the activists were concerned, but the wider social environment in which they were able to exist. At a time when the mainstream media was relied upon to convey their messages to the widest possible audience, my analysis suggests that the women had some success in imparting their antiwar and antiviolence messages to the Australian press, despite the general ambivalence about women and about protesting against war in peacetime. The press coverage of the peace camps was also a way that women were foregrounded as political agents, and this achievement was part of a much wider shift in the way that women came to be represented politically. Through their antiwar and antiviolence activism, women were made "visible as historical actors and as subjects of the narrative" concerned with war and Australian society.[86]

Notes

1. Marilyn Lake and Joy Damousi, "Warfare, History and Gender," in Joy Damousi and Marilyn Lake, eds., *Gender and War: Australians at War in the Twentieth Century.* Melbourne, 1995, p. 3.
2. Ibid., p. 1.
3. Malcolm Saunders and Ralph Summy, *The Australian Peace Movement: A Short History.* Canberra, 1986.
4. Joy Damousi, *Women Come Rally.* Melbourne, 1994; Joy Damousi, "Socialist Women and Gendered Space: Anti-Conscription and Anti-War Campaigns 1914–1918," in Damousi and Lake, eds., *Gender and War.* pp. 254–273.
5. Marilyn Lake, *Getting Equal: The History of Australian Feminism* (Sydney, 1999); Judith Smart, "The Right to Speak and the Right to Be Heard: The Popular

Disruption of Conscriptionist Meetings in Melbourne, 1916," *Australian Historical Studies.* 23, 1989, pp. 203–219. Also see Chris Healy, "War against War," in Verity Burgmann and Jenny Lee, eds., *Staining the Wattle: A People's History of Australia since 1788.* Melbourne, 1988, pp. 208–227; Darryn Kruse and Charles Sowerwine, "Feminism and Pacifism: 'Woman's sphere' in peace and war," in Norma Grieve and Ailsa Burns, eds., *Australian Women: New Feminist Perspectives.* Melbourne, 1986, pp. 42–58.

6. Ann Curthoys, "'Shut Up You Bourgeois Bitch': Sexual Identity and Political Action in the Anti-Vietnam War Movement," in Damousi and Lake, *Gender and War.* pp. 311–341; Lekkie Hopkins, "Fighting to Be Seen and Heard: A Tribute to Four Western Australian Peace Activists," *Women's Studies International Forum.* 22, 1, 1999, pp. 79–87; Joan Williams, "Women Carrying Banners," in Joan Eveline, Lorraine Hayden, eds., *Carrying the Banner: Women, Leadership and Activism in Australia.* Perth, 1999, pp. 16–31.

7. Curthoys, "Shut Up You Bourgeois Bitch," pp. 322–325; Siobhan McHugh, *Minefields and Miniskirts: Australian Women and the Vietnam War.* Sydney, 1993; Barry York, "Power to the Young," in Burgmann and Lee, *Staining the Wattle,* pp. 228–242.

8. Lake, *Getting Equal,* p. 220.

9. Verity Burgmann, *Power and Protest: Movements for Change in Australian Society.* Sydney, 1993, chap. 4; Brendan Cairns, "Stop the Drop," in Burgmann and Lee, *Staining the Wattle,* p. 243.

10. For example, see Suellen Murray, "'Make Pies not War': Protests by the Women's Peace Movement of the Mid 1980s," *Australian Historical Studies.* 127, 2006, pp. 81–94; Diana Pittock, "Women against War and Other Violence," *Peace Studies.* 1985, pp. 16–17, 31–2.

11. I was a participant at and organizer of both camps. For Pine Gap, I was involved with a group of Tasmanian women who traveled to the camp, and prior to our departure, I worked with the local Women for Survival group. After Pine Gap, I returned to Western Australia and brought the message that Women for Survival was keen to have a similar camp there highlighting the U.S. war ships' visits. I then joined with others from the local women's peace group, Women's Action for Nuclear Disarmament (WAND), to make the arrangements for the camp that became known as Cockburn Sound Women's Peace Camp. While acknowledging the influence of my own experiences in the women's peace movement, generally, I write here in the third person referring to the "peace activists" or "Women for Survival" to show that I am referring to the range of views or activities of the wider group (drawn from their literature or press reporting) rather than my own specific experiences.

12. For example, Damousi, "Socialist Women," pp. 254–273.

13. For a more detailed history of the Australian women's peace movement, see Murray, "Make Pies Not War," pp. 81–94. For a history of the British women's peace movement, see Jill Liddington, *The Long Road to Greenham: Feminism and Anti-Militarism in Britain since 1820.* Syracuse, 1989.

14. Ann Curthoys, "Doing It for Themselves: The Women's Movement since the 1970,'" in Kay Saunders, Raymond Evans, eds., *Gender Relations in Australia: Domination and Negotiation.* Sydney, 1992, p. 444.

15. Burgmann, *Power and Protest,* pp. 202–204.

16. "Huge Peace Rallies Jam Major Cities," *West Australian.* 1 April 1985, p. 1.

17. See Stephanie Green, "*Wildflowers* and Other Landscapes,'" *Transformations.* 5, 2002, for comments about the two-week camp held outside the Commonwealth Defense Department Offices in Canberra at the time of the Pine Gap peace camp.

18. As suggested by the acronym, New Zealand is also a member of the ANZUS treaty, but since the mid-1980s, New Zealand has taken an antinuclear stand refusing entry of U.S. war ships and submarines.

19. For descriptions of the Greenham camp, see Alice Cook and Gwyn Kirk, *Greenham Women Everywhere.* London, 1983; Barbara Harford and Sarah Hopkins, *Greenham Common: Women at the Wire.* London, 1984; Margaret L. Laware, "Circling the Missiles and Staining Them Red: Feminist Rhetorical Invention and Strategies of Resistance at the Women's Peace Camp at Greenham Common," *NWSA Journal.* 16, 3, 2004, pp. 18–41; Sasha Roseneil, *Disarming Patriarchy: Feminism and Political Action at Greenham.* Buckingham, 1995. The Greenham Common women's peace camp was to have an enduring influence on the international women's peace movement: David A. Snow and Robert D. Benford, "Alternative Types of Cross-National Diffusion in the Social Movement Arena," in Donatella della Porta, Hanspeter Kriesi, and Dieter Rucht, eds., *Social Movements in a Globalizing World.* London, 1999, pp. 29–30.

20. This is not to suggest that women's understandings did not change over time or that they did not overlap. For example, according to Lawrence Wittner, the "dominant motive" for women's involvement in protests against nuclear war during the 1950s and 1960s was maternalism; while typically not identified as feminist, the activities they engaged in and how they went about it challenged maternalist ideals, and, indeed, some of these women were "swept up by the new feminism" in the 1970s: Lawrence S. Wittner, "Gender Roles and Nuclear Disarmament Activism, 1954–1965," *Gender and History.* 12, 1, 2000, pp. 204, 206.

21. Curthoys, "Doing It for Themselves," p. 444.

22. Lauri Buckingham, "Mobilisation of Women Leading up to the Pine Gap Action," in Michael J. Roache and Anne Curthoys, eds., *Not the Bicentennial: A Collection of Essays on Australian History, Sociology and Politics.* Sydney, 1988, pp. 24–43.

23. Women for Survival, *Pine Gap Camp Handbook,* 11–25 November 1983, p. 2.

24. WAND, "The USA Presence in Australia," (flier), c. 1984.

25. Women for Survival, *Pine Gap Camp Handbook,* p. 2; Women for Survival, *Sound Women's Peace Camp Information Kit,* 1–15 December 1984, p. 4; Lee O'Gorman, "Australian Women's Unique Action for Peace," *Women of the Whole World.* 2, 1984, pp. 14–15.

26. For the presence of U.S. sailors in Fremantle, see Murray, "Make Pies Not War."
For views of U.S. sailors during Second World War, see Kate Darian-Smith,
"Remembering Romance: Memory, Gender and World War II," in Damousi and
Lake, *Gender and War*, pp. 117–129; Marilyn Lake, "Female Desires: The
Meaning of World War II," in Damousi and Lake, *Gender and War*, pp. 60–80.

27. Buckingham, "Mobilisation of Women Leading up to the Pine Gap Action,"
p. 41.

28. For example, articles about the peace camps in feminist newsletters such as
Perth's *Grapevine*, various PND (People for Nuclear Disarmament) newsletters,
the environmentalist and antinuclear magazine *Chain Reaction*, and the Left
newspaper *Direct Action*. However, not all feminists were supportive of the
actions that the peace activists were taking. An editorial in the Australian
feminist literary journal *Hecate*, argued that the perspectives of women's peace
actions were "essentially . . . middle class" and that the Pine Gap camp would
be "ineffectual . . . at challenging Hawke's recent 'cementing' of the US military
alliance." Editorial, *Hecate*. 9, 1/2, 1983, p. 4.

29. Kristine Anderson, "Women's Peace Camp—Cockburn Sound, Western
Australia," *Union of Australian Women Newsletter*. March 1985, pp. 7–8;
O'Gorman, p. 15.

30. For example, for the Pine Gap camp, two-meter-high banners depicting life-size
figures of those who could not be there were made in a project entitled "Double
our Numbers" and organized by Alice Springs Women for Survival. The kilometer-
long stream of hundreds of banners was carried on the first day as the women
marched up the road to the gates of Pine Gap. Gillian Fisher, "Remember Pine
Gap," *Burn: Proud to be Different*. November 1993, pp. 30–33.

31. Greenham Common women very successfully promoted their activities through
similar methods: Julie Emberley and Donna Landrey, "Coverage of Greenham
and Greenham as 'Coverage,'" *Feminist Studies*. 15, 3, 1989, pp. 491–492.

32. *Western Australian Parliamentary Debates,* Legislative Council, 22 August 1984,
p. 1083.

33. My discussion here is focused on the printed press, but the peace activists also
worked with others in television and radio.

34. Collectives were established at both women's peace camps to facilitate the running of the camp and to ensure that it occurred in what were considered to be
democratic ways. There were collectives that had responsibility for legal matters,
healing, police liaison, garbage, water, security, and child care, as well as for the
media. For a description of the internal workings of one of these collectives at
another later Australian women's peace camp, see Mary Heath, "Peace, Protests
and Police: Police Liaison at a National Women's Peace Action against Australian
Militarism," *Alternative Law Journal*. 20, 6, 1995, pp. 291–293.

35. Women for Survival, *Sound Women's Peace Camp Information Kit*, p. 12.

36. For discussion of another smaller contemporary collective and the difficulties
(as well as the joys) it encountered, see Suellen Murray, *More than Refuge:
Changing Responses to Domestic Violence*. Perth, 2002, pp. 67–76. For discussion

of collectives at another women's peace camp, see Peregrine Schwartz-Shea and Debra D. Burrington, "Free Riding, Alternative Organization and Cultural Feminism: The Case of Seneca Women's Peace Camp," *Women and Politics*. 10, 3, 1990, pp. 1–37.

37. Women for Survival, *Sound Women's Peace Camp Information Kit*, p. 12.

38. Bruce Stannard, "Peace Takes a Back Seat to Feminism," *Bulletin*. 29 November 1983, p. 27.

39. Lorraine Brown, "Male Ban in Peace Protest," *Sunday Independent*. 11 November 1984, p. 15.

40. Various Western Australian female journalists expressed their view on the issue of women-only journalists in the peace camp, for example, "We Get Women's Support," *Daily News*. 12 November 1984.

41. Cyril Ayris, "In a Confrontation, the Eyeballs Have It," *West Australian*. 4 December 1984, p. 1; "Peace by Force" (editorial), *West Australian*. 5 December 1984, p. 8.

42. "Sound Peace Camp: On the Inside," *Sound Telegraph*. 12 December 1984, p. 3. For similar sentiments in the press reporting of Greenham Common, see Roseneil, *Disarming Patriarchy*, pp. 130–132. Such negative portrayals of lesbians (and other feminists), however, were not unique to the women's peace movement: Debra Baker Beck, "The 'F' Word: How the Media Frame Feminism," *NWSA Journal*. 10, 1, 1995, pp. 1–26; Deborah L. Rhode, "Media Images, Feminist Issues," *Signs*. 20, 3, pp. 1–26.

43. Waller (cartoon), *Western Mail*, 1 December 1984. Also see Alan Langoulant (cartoon), "Paradise Lost," *Daily News*. 7 December 1984, p. 52.

44. "Liberals Say Peace Camp a Grubby Event," *West Australian*. 19 November 1984; "Protestors Unfeminine" (letter to the editor), *Centralian Advocate*. 16 November 1983. Similar references were made to the lack of cleanliness of the women and their campsites at Greenham Common, see Roseneil, *Disarming Patriarchy*, pp. 130–132.

45. Barbara Brook, "Femininity and Culture: Some Notes on the Gendering of Women in Australia," in Kate Pritchard Hughes, ed., *Contemporary Australian Feminism 2*. Melbourne, Longman, 1997, p. 107.

46. For portrayals of "maternalist" peace activists, see Sally Abbot, "The Sound of Peace at Cockburn," *Daily News*. 3 December 1984, p. 3; for "feminist" peace activists, see Kim Murray, "Into Battle with a Frenzied Beat," *Daily News*. 7 December 1984, p. 4; Cyril Ayris, "Peace and Harmony," *West Australian*. 8 December 1984, p. 16. The latter, particularly, were portrayed unflatteringly in cartoons by an unknown artist, *Northern Territory News*. 19 November 1983, p. 7; Mariusz(cartoon), *West Australian*. 7 December 1984, p. 8.

47. Emberley and Landrey, "Coverage of Greenham and Greenham as 'Coverage,'" pp. 485–498; Roseneil, *Disarming Patriarchy*, pp. 130–132, 170–172.

48. See, for example, letters to the editor expressing disquiet about the "violent" protest actions in D. C. Airey, "Call on Government to Act," *West Australian*, 6 December 1984, p. 6; M. R. Barker, "In the Name of Peace," *Daily News*. 14 December 1984, p. 19.

49. Laware, "Circling the Missiles and Staining Them Red," p. 28.
50. Ibid., p. 30.
51. From a press release issued by Women for Survival and cited in Kim Murray, "Women storm gate—40 arrested," *Daily News.* 6 December 1984, p. 1.
52. For example, see "Protesters Leaving," *Northern Territory News.* 17 November 1983, p. 1; "WA Protest: Women Held," *Australian,* 7 December 1984, p. 2; Cyril Ayris and Peter Denton, "Women Leaving Peron Protest," *West Australian.* 8 December 1984, p. 2; "Peron Protest Ends," *Daily News.* 14 December 1984, p. 2.
53. Michael Barnard, "Whose Zoo at Pine Gap?," *Northern Territory News.* 26 November 1983, p. 7.
54. Jill Bottrall, "It's D-Day at Pine Gap," *Centralian Advocate.* 11 November 1983, pp. 1–2.
55. "Grandmother Aggrieved" (letter to the editor), *Daily News.* 7 December 1984, p. 22.
56. "Date for Moscow" (letter to the editor), *Daily News.* 7 December 1984, p. 22.
57. Also see Mariusz (cartoon), "Just Ignore Them and They'll Go Away," *West Australian.* 16 November 1983, p. 8.
58. Danielle Robinson, "Protestors Take to Perth Streets," *Australian.* 9 November 1984, p. 24.
59. Emberley and Landry, "Coverage of Greenham and Greenham as 'Coverage'," p. 492.
60. Sidney Tarrow, *Power in Movement: Social Movements and Contentious Politics.* 2nd ed. Cambridge, 1998, p. 116.
61. Joan Williams, "A Significant Event: Cockburn Sound Women's Peace Camp, 1–15 December 1984," *Papers in Labour History.* 10, 1992, pp. 25–34; Terri Seddon, "Pine Gap Women's Camp: Disarmament and Power Relations," *Peace Studies.* December 1984, p. 26.
62. Amanda Buckley, "Invasion of Pine Gap: Feel the Heat (40ºC), Face 111 Karen Silkwoods Head On. Call in the 'Copter,'" *Sydney Morning Herald.* 14 November 1983, p. 1.
63. This action was reported widely in Australia and internationally: "111 Militant Women Seized at Australia Base," *New York Times.* 14 November 1983, p. 5.
64. "Pine Gap Crusaders Arrested," *Age.* 14 November 1983, pp. 1–2.
65. Jill Bottrall, "On the Road to Pine Gap . . . ," *Centralian Advocate.* 16 November 1983, p. 1.
66. Senator Gareth Evans, the then federal Attorney-General, ordered an inquiry into the protestors' ill-treatment; Human Rights Commission, *Report 20: Complaints Relating to the Protest at Pine Gap.* Canberra, November 1983.
67. Sue Murray, personal journal, Pine Gap Women's Peace Camp, 15 November 1983. Another account of this action is found in Margaret Somerville, *Body/Language Journals.* Melbourne, 1999, pp. 33–35.
68. Norm Lipson, "The Battle of Pine Gap," *Daily Telegraph.* 16 November 1983, p. 1.
69. *Sydney Morning Herald.* 16 November 1983, p. 1. Also see Simon Balderstone, "Arms and the Women," *Age.* 16 November 1983, p. 1; "Pine Gap Protest Bubbles

up Again," *West Australian.* 16 November 1983, p. 2; Jill Bottrall, "Chaos as Security Gates Fly," *Centralian Advocate.* 16 November 1983, p. 2.

70. For example, unknown photographer, "Going, Going, Gone . . . ," *Australian.* 16 November 1983, p. 1.

71. For example, see "Hundreds Arrested in UK Demonstrations," *Northern Territory News.* 16 November 1983, p. 3; "Cruise Missiles at UK Base," *Mercury.* 16 November 1983, p. 5.

72. Sylvia Monk (letter to the editor), "She Won't Be Right," *Centralian Advocate.* 23 November 1983.

73. Nicholson (cartoon), "Why Don't the Media Report the Real Issue . . . ," *Age.* 21 November 1983, p. 13.

74. Cyril Ayris and Pilita Clark, "Vandals Spoil Protest Gains," *West Australian.* 4 December 1984, pp. 1–2. On page 2, where the front-page article concluded, it was headed "'Fringe' Artists Spoil Protest."

75. For example, see Jan Mayman, "Women Protesters March on Base," *Age.* 4 December 1984, p. 3; "Women Demanding Disarmament March at Point Peron: Peace Group Calls on Scholes," *Mercury.* 4 December 1984, p. 12.

76. Sue Murray, personal journal, Sound Women's Peace Camp, 7 December 1984. For another account of this action: Williams, "Women Carrying Banners", pp. 29–30.

77. Cyril Ayris, "Assault on Naval Gate Lands 75 in Gaol," *West Australian.* 7 December 1984, p. 1.

78. Murray, "Women Storm Gate," p. 1.

79. This criticism included discussion in the "Peace by Force" (editorial), *West Australian.* 5 December 1984, p. 8; "WA Protest: Women Held," *Australian.* 7 December 1984, p. 2.

80. Diana Callendar, "Women Take Hassell at Face Value," *West Australian.* 7 December 1984, p. 2.

81. Cyril Ayris, "Peace Camp Raided by State Police," *West Australian.* 13 December 1984, p. 3.

82. W. Hartley (letter to the editor), "Motives Praised," *West Australian.* 12 December 1984, p. 8.

83. Ailsa Ruse (letter to the editor), "Points on the Protest," *West Australian.* 12 December 1984, p. 8.

84. "50 Arrested as Women Storm WA Navy Base," *Age.* 7 December 1984, p. 1.

85. Emberley and Landry, "Coverage of Greenham and Greenham as 'Coverage,'" p. 487.

86. Lake and Damousi, "Warfare, History and Gender," p. 1.

CHAPTER 9

The War at Home: Toys, Media, and Play as War Work

Karen J. Hall

> Every commodity reproduces the ideology of the system that produced it: a commodity is ideology made material.
>
> John Fiske[1]

War and armed aggression have ancient roots, but industrialization and globalization have heightened the intensity of militarism's effects on civilian society. Consumer lives in the industrialized world are saturated with products directly or indirectly tied to the military industrial complex. Those who live their lives on the margins of the industrialized world cannot escape militarism either; such places, in most cases former colonies or current economic zones of interest for industrialized nations, have become the dumping grounds of new and used popular culture as well as first- and secondhand military hardware. A child wearing a Transformers t-shirt is just as likely to live in Colombia as in the Philippines, the United States, or Ghana. Technology and globalization have increased the rate and range of exchange from one location to another. The toys and images that flood one market are rapidly deployed globally, making war toys an issue for the entire global village.

Children form their moral perspectives at a very young age. Toys and play are crucial to the development of their understanding of the world. When war toys and aggressive action figures dominate children's toy boxes and playtime, war toys become an influential source that teaches children some of the core values of militarism. Physicians for Global Survival (PGS), a physicians' peace activist group founded in the early 1980s, identifies the four most-prominent moral lessons that war toys teach children as the following:

- War is a game, an exciting adventure.
- Killing is acceptable, even fun.
- Violence or the threat of violence is the only way to resolve conflicts.
- The world is divided into "goodies" and "baddies," where the bad guys are devoid of human qualities and their destruction is desirable.[2]

War play that rehearses stories in which "bad" people, who seek to control the world, are defeated by "good people" teaches children that weapons and war are sources of power and are necessary to deal with "evil." Militarist media feels compelling, sets audience expectation and excitement levels, is all but impossible to avoid, and defines fun as action- and adrenaline-based. War entertainment currently sold to global consumers is not responsible for creating militarist values, but it goes a long way in reinforcing the lessons that help militarism maintain its hegemony.

The encroachment of militarism into the lives of consumers and the seductive dangers it poses are the explicit topics of director Joe Dante's 1998 Hollywood feature film, *Small Soldiers*. This satire of late twentieth-century U.S. militarist-laced consumerism was ideally built for the market environment it critiques: the movie wittily partakes in the very elements it satirizes. *Small Soldiers* sells its antimilitarist theme in packaging that highlights the action-based narrative and explosive spectacle that captivates a broad spectrum of the mainstream viewing audience. Thus, it should come as no surprise that the film itself faced protest for glamorizing violence. However, it is my belief that teaching the world's children to play with the dangerous double-edged sword of militarism's toys with a knowing and satiric awareness is the best defense activists have to offer. *Small Soldiers* can offer a model for new forms of intelligent engaged play that promotes learning about the dangers of war and militarism in our world, making the film and its toy tie-ins entertaining and instructive products for children and adults to explore.

Dangerous Incursions into Toyland

From the time their first T-shirt is snapped beneath the crotch of their diapers, it is possible to surround children in the industrialized world with images of war. Tanks, fighter jets, and soldiers are available to decorate the bumpers on infants' cribs and the mobiles over their heads so that some of the first images their minds absorb directly support militarism.[3] Peace-minded parents can only shelter their offspring for so long; as soon as children mix with peers or gain access to media, the images of war begin to flood their conscious and unconscious minds. Militarist consumer products become increasingly interactive as children age; until one day, the camouflage musical mobile's place is taken by a first-person shooter game so realistic that the military uses it to train their own *real* soldiers. In between these two extremes, GI Joe, Hot Wheels, no-name plastic soldier sets, Hometown Heroes, Transformers, and this year's hot new weaponized toy are all ready, willing, and able to stock children's playtime arsenals.

Concerned citizens have spoken out against turning children into soldiers by means of war toys since the nineteenth century. Toy guns were the earliest focus of this attention as demonstrated by the 1933 *Washington Post* headline "Pacifist Mamas Ban Toy Soldier Wars."[4] GI Joe, the U.S. toy made infamous to toyland pacifists by its long run and vast array of product tie-ins, was met at its inaugural New York Toy Fair in 1964 by Parents for Responsibility in the Toy Industry, who carried signs that read "Toy Fair or Warfare."[5] From 1964 to 1968, a particularly successful era for the GI Joe product line, activists encouraged parents to purchase toys that would stimulate children's creativity, like the Swedish brand Lego blocks, rather than Hasbro's toy soldier, which would train them to imitate national militarist ideology.[6] While individuals were willing to speak out against violent toys, the most well-known organization in the United States dedicated specifically to promote demilitarization and antiviolence in the toy industry was the Lion and Lamb Project, founded in 1994.[7] For almost ten years, Lion and Lamb produced an annual list of the "Dirty Dozen," the year's most violent toys as well as a list of twenty creative nonviolent toys. Although the Lion and Lamb Project is no longer active, its founder, Daphne White, has published guidelines for how their lists were developed in an 2004 issue of *Mothering* magazine so that parents have some rules to guide their toy purchases.[8] Supported by little else other than this 2004 magazine article, parents are for the most part on their own when trying to negotiate healthy boundaries and practices for their children's play.

Despite the fact that militarism continues to offer children of all ages toys with increased complexity and spectacular effects, the strategies and alternatives provided by peace-minded activists have not advanced apace, and no organization in the United States has stepped up to fill the place of the Lion and Lamb Project. Thus, the issue of war toys is brought to the public's attention perhaps once annually, during the winter holiday toy-buying season, and those who protest toy and play violence have a hard time being taken seriously. Much like the healthy new-age toy store in Joe Dante's *Small Soldiers*, the Inner Child, toy marketers have not developed nonviolent products that can compete in today's marketplace. When I think about the

"No War Toys" movement in my own community, I am reminded of Nancy Regan's "Just Say No" antidrug campaign. For many years the local peace and justice community held an alternative toy fair in December, but the goods were as boring and lackluster to contemporary children as those in Stuart Abernathy's the Inner Child toy store. Saying no to addiction, whether chemical or militarist, is indeed crucial, but there is no "just" about it—unless of course we mean to call upon the adjectival form of "justice" rather than the adverb meaning "only." Increasing the strength and status of justice worldwide would indeed do much to tame the violent scenarios children act out with their militarist toys. However, there is nothing simple or isolated about even imagining such an undertaking. Far more simple and lucrative than to create a new world of justice or resist the present world's many injustices is to imagine, develop, and market new products that exploit the injustices and dangerous power differentials plaguing our world.

Friend or Foe: Small Soldiers Toy Tie-Ins and Protests of Violent Toys

Developing a fresh new line of war toys is just what Hasbro did to accompany the release of *Small Soldiers*. Although licensed characters seem to have taken over stores' toy shelves in the past 10 to 20 years, connecting licensed characters to children's toys is as old in the United States as licensed characters are. From Dick Tracy toy guns to *Star Wars* action figures and beyond, children have played with media-driven narratives throughout the twentieth century. The later years of the century and the opening of the twenty-first century, however, have seen ever more aggressive marketing and tie-in campaigns. These trends helped to lead the toy historian Howard Chudacoff to the claim that "commercial toys have almost completely colonised children's free time."[9] It is difficult to grasp the extent to which consumerism has laid claim to childhood. Experts estimated that in the year 2000, children between the ages of 4 and 12 were directly responsible for $170 billion in spending in the United States alone.[10] Media critics and children's advocates worry that the colonization of children's time has been accompanied by the colonization of their imaginations. As with any colonial project, cultural forces do not gain control of even intangible regions like children's imagination without at least a show of violence.

In the case of *Small Soldiers,* the show of force came in the form of massive numbers. Sixty licensees released what has increasingly become the standard list of related products in tandem with *Small Soldiers,* including a soundtrack, video games, action figures, miniature cars, bedsheets, clothing, fast-food toy premiums, and trading cards.[11] While the movie was in vogue, its images and characters could rule a child's universe. The July release date for the movie made the toy line the number one boys' action toy for the summer and earned the movie a berth in *Business Week's* product-placement hall of fame—recognition that would delight the film's fictional CEO, Gil Mars, but that audience members in on the satire would understand ironically.[12]

In their work *Consuming Children: Education-Entertainment-Advertising*, Jen Kenway and Elizabeth Bullen highlight how the market is able to communicate directly with children, decreasing the significance of family, school, or church as central social structures in children's lives. This situation is even more alarming when one considers the trends in militarized entertainment violence and the glorification of war in popular culture. The U.S. Federal Trade Commission has criticized companies and advertisers for marketing R-rated films to children.[13] For example, the R-rated film *Matrix* had a corresponding teen-rated video game. Although less severe an age gap, the toy premiums associated with *Small Soldiers* and distributed with children's meals at Burger King were suitable for the two-to-eight-year-old market group while the movie was rated PG-13.[14] Burger King did move their television advertising campaign from Saturday morning, a time slot that operates as prime time for children, to the adult prime-time hour and offered alternative toys. However, enthusiasm for a movie that had been deemed excessively violent by the Motion Picture Association still reached the under-13 audience. When this marketing information is taken into account along with the facts that Burger King had a final edit of the movie and Hasbro helped develop the characters, the media's role in promoting violence and militarization in children's culture is damning.[15]

The case against allowing children to play with a media matrix like *Small Soldiers* is not to be taken lightly. Like many cultural critics, I too am alarmed by the degree of media saturation represented by product launches like that of *Small Soldiers* in our own and in our children's lives. Along with many critics of postmodernism, I believe that the saturation of images has led to an aestheticization of reality and a distancing of affect.[16] In such a world of images, viewers are invited to grow numb to the felt experience suffered by those represented. A study of children's attitudes toward war toys that collected interviews and data in 1985 and again in 2002, showed a "dramatic decrease in statements that reflect awareness of the horrors of war and terror and a pacifist attitude."[17] In 1985, children who took part in the survey referred to the horrors of nuclear war, acknowledged the vast number of people affected in war, and voiced such vivid concerns as, "I don't like people shooting at each other, and then lying there with big wounds, still living for a little bit, and then dying." By 2002, children interviewed only voiced bland and very general antiwar sentiments: "I think it's naff to fight, because mostly you kill everybody" or "Sometimes war is senseless."[18] As signifiers are heaped upon signifiers, any "authentic" understanding of the real becomes more unreachable, so that no source is to be trusted and eventually audiences lose interest in knowing, especially when imagining is easier and more entertaining. Bringing an experience of war home to viewers is one of the critically important aspects of *Small Soldiers*. The blockbuster movie invited viewers to intimately imagine the power of the violence lying (temporarily) latent in their toys; once the Commando Elite punch their way out of their boxes in this film, the home front becomes quite literally a warfront.

The critic Roger Ebert claimed that "*Small Soldiers* is a family picture on the outside, and a mean, violent action picture on the inside."[19] In spite of corporate sponsorships and back-end control, director Joe Dante, along with the cast of writers involved in the project, was able to create an intelligent, witty, and fun movie that challenges the very aestheticization and cultural numbing trafficked in war-entertainment. After all, this story's battle is waged by the coolest toys any kid could ever imagine, and it takes place not in some far and distant land but in a typical upper-middle-class U.S. suburban neighborhood in homes many U.S. viewers can identify with as their own. Because *Small Soldiers* brings the violence in war toys to life in its audience's own living space, the movie invites mainstream audiences, at least for a moment, to feel the destructive power that is taking over children's imaginations. Without shaming or berating consumers who enjoy today's militarized entertainment spectacles, *Small Soldiers* manages to convince even little Timmy Fimple that the best birthday present is not a cool toy soldier but new clothes.

The Smell of Satire in the Toy Room

The media text that is *Small Soldiers* invites children to play and learn for themselves why some adults prohibit war toys. Satire requires trust in one's audience: if viewers read the text straight, it will offer morally unsound examples. However, if audiences learn to play with the text as it was designed, in this case as a humorous satire with a useful moral, they will discover that satiric works offer audiences lessons in interpretation that strengthen their life-reading skills. If consumers are going to learn to negotiate the intricacies of militarism, they are going to need to have well-developed powers of interpretation. The simplistic good-versus-evil moral lessons media, school, and government continue to offer are not going to prove helpful, and new skills will have to be attained. The double-edged dual nature of satire is an important element to learn to negotiate and *Small Soldiers* is an enjoyable primer.

The movie opens with the parody of a commercial for Globotech Corporation, "long recognized as the world-wide leader in high tech weaponry." Under the leadership of CEO Gil Mars, Globotech is expanding its interests into "tomorrow's most exciting market sector. Introducing advanced battlefield technology into consumer products for the whole family." Even if the commercial's dialogue does not have great meaning to young viewers, they will be savvy enough to know that the typical advertisement does not combine images of military tanks, missiles, and warships with the style of clichéd family scenes—on the beach, in the living room, with a new baby—that this commercial presents.[20] Thus, from the opening minutes of the film, viewers are introduced to the idea that the military and home spaces are combining in strange and unexpected ways. The binary opposition between the military and the domestic is dangerously resolved via dialectical synthesis, and two new entities are created: the domesticated-military and the militarized-domestic.

The first four characters introduced are CEO Gil Mars, his female assistant, and two bumbling toy developers. In addition to being able to recognize the disconnect between military hardware and domestic space, most children will also be sophisticated enough to recognize who the formulaic good guys and bad guys are: the boss who has fired all the people, who would be sitting around the large wooden table, is the bad guy and the two toy makers are the good guys. However, the narrative begins to play immediately with viewers' identifications; the bad guy is tired of commercials that lie to consumers, promising more than they can ever deliver. Mars also knows that kids do not want to be tricked into learning by their toys. Even though Irwin Wayfair is the more gentle good guy, his concept for the Gorgonite toy line that helps children learn and do research shows his nerdy naiveté and questionable understanding of his target market. Mars seems to know what children want and insists that his company produce cool toys with batteries included that are capable of lasting a lifetime. Again, from the very opening of the film, the audience is introduced to a dynamic that is significant to the entire text: the ability to shift identification with and expectations of characters. Mars, who seems like a bad-guy boss, leaves the brief scene speaking on behalf of kids and their desires. Whether or not young audience members will be able to balance the side of Mars they are bound to like with the side that is dangerously cavalier will depend on the child. For example, Mars states:

> We can make a missile that can hunt one unlucky bastard 7,000 miles away and stick a nuclear warhead right up his ass. I don't think we're going to have a problem with this [the technology to make the toys interactive].

It is difficult to resolve the interests of giving some children what they want while using technology with the same capacity to possibly kill other children who live those 7,000 miles away. A majority of viewers will have seen that he has a bad-guy and a good-guy side and will have experienced some shift of identification with him. This initial shift and those that follow introduce viewers to the complex and unstable breakdown of binary opposition. Contrary to the dominant war narrative that teaches viewers that the bad guys are inherently evil and will ultimately, after great sacrifice, be defeated by the good guys, *Small Soldiers* complicates almost every character's intentions and moral standpoint.

The most startling shift in audience identification occurs once it is clear that the Commando Elite, the toy stand-ins for the U.S. military, are the bad guys. Viewers who have been in synch with the movie's satiric positioning from the opening segment will have expected the Commandoes' violent behavior, but younger and more naïve viewers used to seeing U.S. soldiers convey redeeming characteristics may be shocked as the Commandoes bind and take the Fimple children hostage, drug Mrs. Fimple, launch flaming tennis balls into a civilian home with no regard for collateral damage, and mutilate an extensive collection of female fashion dolls. This is not the kind of viewer identification for which popular media has prepared U.S. children. Siding with the Gorgonites who are mutant, monstrous freaks who fight (albeit reluctantly) against U.S. soldiers is the stuff of propaganda, not mainstream, summer blockbusters made for the family-viewing audience. The irony of a mainstream product delivering such a marginal critique is compounded by the fact that the film was produced by Dreamworks, a motion picture studio cofounded by Stephen Spielberg, who, due largely to the making of *Saving Private Ryan* is favored in the U.S. populist view as the most renowned and reliable teller of war tales alive today. The movie subverts populist knowledge and positioning by making Mr. Fimple, a likely fan of Spielberg's work given that he claims his favorite war is World War II, one of the film's humorous fall guys rather than a valued consumer patriot. These inversions of viewer expectations enhance the carnivalesque aspects of the film. During carnival, the low masquerades as high and the high as low, but there are no costumes in *Small Soldiers*; instead, masquerade has been dropped, exposing the moral depravity of the high and lifting the low to its more rightful position.

The pure evil of the Commando Elite was what threw off the moral balance of the film for Roger Ebert, leaving him feeling as though the movie "didn't tell me where to stand—what attitude to adopt."[21] I am unsure how to reconcile Ebert's moral confusion with Scott Rosenberg's claim in *Salon.com* that the film is marred by its "fairly heavy-handed message about The Evil

That Is War Toys and a cautionary invocation against corporate domination of the entertainment market," both of which he recommends ignoring in order that audiences may "simply enjoy the filmmakers' skill at creating carefully contained mayhem in a microcosm."[22] The movie makes it quite clear that the Commando Elite are ruthless and treacherous. After all, it is they who destroy two suburban homes and endanger the men, women, and children in two families just because they have been programmed to hunt down and kill the Gorgonites. One might argue against blaming the toy soldiers due to the fact that, unlike actual military personnel who commit atrocities, the Commando Elite never had the power of free will because they can only act as their program dictates. *Small Soldiers* disrupts the messages conveyed by typical war toys without shaming or blaming anyone. The movie makes it clear to viewers of all ages that killing is not fun when you are the target and that fighting seems like a terrible option to resolve conflict when you are the weaker combatant positioned to lose all. The movie turns the tables of morality so that its U.S. audience can imagine itself not as the righteous victim of an evil enemy's violence, as it most typically does, but as complicitous perpetrators whose militarist investments at last cause us to shoot ourselves in the foot.

Although the only logically sound response to the Commando Elite is to destroy them, Gil Mars has a different plan. After paying huge sums of money to the Fimple and Abernathy families in order to buy their silence and prevent lawsuits, Mars holds the following conversation with his two toy developers:

> "What were we charging for these things?" [asks Mars as he looks at the severed head of Chip Hazard].
> "$79.95."
> "Tell you what, add a few zeroes to the end of that number and get in touch with our military division. I know some rebels in South America who are going to find these toys very entertaining" [says Mars as he returns to his helicopter to be whisked from the scene].

The satire of *Small Soldiers'* is comprised of more than simple inversion. The film pokes explosive fun at the lie that the domestication of militarism will benefit the private sector consumer. Gil Mars, the very man who claims that he is tired of commercials that make promises they can never deliver, is the film's spokesperson for the hucksterism that is militarism. Globotech professes to turn swords into plowshares when, in reality, the narrative exposes how defense industries resell military products to the domestic consumer in order to increase their profit—turning swords into only slightly duller swords for domestic consumption. The private sector increases the defense industry's market, driving down the cost impact of research and development, and increasing sales. For the defense industry, consumer sales are a winning proposition; civilian consumers get to fund the defense industry first with their federal taxes and again with their purchases, paying dual support for war whether they are aware of it or not. The resale of military goods is especially prevalent in high-tech products like computer games and Global Positioning Devices.[23]

The destruction of the Fimple and Abernathy homes is an exaggerated representation of the downside of the conglomeration of industries under the military umbrella, yet it does use exaggeration effectively to make the case against the domestication of militarism tangibly clear. Viewing the circumstances of the destruction of the Fimple and Abernathy homes makes *Small Soldiers* a very different narrative engine than the similarly destroyed and militarized domestic space that was the Forward Command Post. Toy manufacturer Ever Sparkle Industrial Co. Ltd. produced this bombed-out dollhouse for the 2002 toy season.[24] Sold through JC Penney, Toys "R" Us, and eToys, at the cost of U.S.$45, this toy came in a 75-piece set that included one action figurine in military combat gear, multiple toy weapons, an American flag, tables, chairs, and a three-room house with shattered walls and bullet-pocked plaster. The Forward Command Post was not an ironic toy. Like the majority of war toys produced that simulate twentieth-century combat, the toy aesthetic was based on realism. Children were no doubt meant to learn about history and strategy from this mutilated dream house. Thus, the style of play most likely to take place around this toy was one commanded by the governments and industries of the industrialized world: a play narrative where government forces regretfully occupy civilian buildings in order to plan strategic attacks and secure an area's safety. The style of play encouraged by *Small Soldiers'* products work off a narrative base that exposes both the lie and the cost of this dominant pro-militarist narrative.

In *Small Soldiers*, children are not innocent victims in need of paternalistic protection but are actually invited to see how they may hold some responsibility for the mayhem created by entertaining themselves with these pro-militarist narratives. The film also illustrates the tensions between giving children what is "cool" and struggling both personally and financially to maintain a moral antiwar and antiwar-toy parenting stance. Alan Abernathy means well when he brings a full set of Commando Elite and Gorgonite toys into his father's toy store, The Inner Child, but he also knows he is going against his father's strict policy against war toys. Alan is bored and embarrassed by his parents' new-age lifestyle. As a new kid in school, he knows the cool guys drive motorcycles and is not above cultivating a bad-guy persona even if he knows it is not truly who he is. When he sees the chance to bring cutting-edge technological toy soldiers into his father's struggling shop, he convinces Joe, Globotech's driver, to "lose" a full set of toys. Although I do not agree with the Christian commentator Israel Canlapan's view of the film, his list of grievances do all take place:

> There are ideas about setting a moribund business on fire to collect insurance, conniving to retain merchandise in violation of proper business ethics, and paying people enough money to silence their indignation. If that weren't enough, this movie is filled with many negatives of present society: a cut-throat business magnate who has little regard for moral values, a co-worker who undermines his partner's trust in the name of business success, a trouble-making kid in a dysfunctional family, and a neighbor who rudely violates his neighbor's property.[25]

Alan does indeed sarcastically suggest that the only way his father's toy store will make money is if he burns it down, and this is how he justifies his unethical decision to take stock bound for another store and break his father's "no war toys" policy. What Canlapan fails to appreciate is that Alan learns his moral lesson without a harping moralizing authority figure shaking a finger in his face. Life unfolds a world of hurt that teaches Alan that war toys are hazardous to all he cares about—his home, family, girlfriend, and even his own life. All of the bad behavior that has offended Canlapan is punished by the film's end, and it is the Commando Elite, the metaphorical face of unchecked militarism, who deliver the discipline and punishment.

Quite unwittingly, the two toy developers, Irwin Wayfair (perhaps the only character who, from start to finish, is a *way fair* good guy) and Larry Benson, harness the defense industry's will to control life and death when they put artificial intelligence into their electronic toys, making *Small Soldiers* a modern-day *Frankenstein*. Trapped in a system that demands innovation and excitement yet refuses to make the necessary expenditures for product safety and testing, Wayfair and Benson create the monsters that are the Commando Elite. In order to keep their jobs (Mars has gutted the labor force from the company that was Heartland Playsystems), Benson and Wayfair have three months to bring the Commando Elite to the market-place. Hooked into the database of Globotech's military division, Benson orders military surplus microchips to animate the toy line, thus setting in play the catastrophe that will wreck the two suburban homes. Benson is a morally suspect character; however, like Alan, he is simply trying to survive in a competitive toy industry where action and violence sell and educational inner-child-strengthening toys earn you a place on the unemployment line. To Benson's credit, once he and Wayfair learn that something has gone wrong with their inventions, they learn how to dismantle the toys and go to Alan's assistance.

The key to the Commando Elite's destruction is found in property-destroying, war-loving, technophilic, obnoxious neighbor Phil Fimple's house. The embattled suburbanites turn on all of the electrical equipment in Fimple's house and then short-circuit the transformer on the power line outside, creating an electric burst that shuts down the Commando Elite and offers an example of a story in which the master's tools really can dismantle the master's house. With the Commandoes thwarted, Alan is free to feel the remorse he has brought on himself by knowingly breaking rules. Because he has felt the danger militarism brings into domestic spaces and has witnessed its destruction, his understanding is more likely to remain with him. Having lived through the fear and life-threatening reality of combat, war is less likely to appear as a child-appropriate game. Thus, in the end, Alan has taught himself to understand and share his father's rejection of war toys.

While I am confident that Alan will remember the lesson he learned about militaristic war toys, I cannot be as confident that the *Small Soldiers* viewing audience will be equally affected, especially when they are exposed to aisle upon aisle of *Small Soldiers* products. Given that I have claimed that adequately negotiating the interpretative challenges of a satiric text is difficult and requires a skill set that few consumers have honed, in all likelihood, once the movie narrative moves from the movie screen to the playroom, dominant promilitarist narratives will have their opportunity to reclaim any territory lost to *Small Soldiers'* satiric critique. When Chip Hazard and Archer mix it up with Teenage Mutant Ninja Turtles, Transformers, Power Rangers, and the like, old patterns will override the inversions practiced in the movie. However, when children replay their favorite scenes from *Small Soldiers*, the film's text at least opens the opportunity that they will create alternative storylines, oppositional muscle memory, and ironic literacy for themselves.

Let the Games Begin: Reclaiming and Relearning Play

When houses begin to fill with *Small Soldiers* products, parents need not necessarily worry. Satire is a taste that children have the ability to acquire. I want to challenge adults who sound like Roger Ebert when he opines that:

> what bothered me most about *Small Soldiers* is that it didn't tell me where to stand—what attitude to adopt. In movies for adults, I like that quality.

But here is a movie being sold to kids, with a lot of toy tie-ins and ads on the children's TV channels. Below a certain age, they like to know what they can count on.

How do adults know that children want texts that have the degree of moral certainty and closure that Ebert suggests they do? Just because adults like Ebert do not like to play with possibilities at the end of a story does not mean that children will not enjoy play of this sort. The play theorist Brian Sutton-Smith argued that ambivalence is one of the elements that differentiates play from reality.[26] Play oscillates between mastery and discovery of the child's environment and self-detachment and distance from that environment as well as between tension and relief. There is always the danger that one pole will overpower the other and that play will be ruined. Ambivalence is one of the crucial terms of play that children learn to negotiate, at times tipping the balance toward tears and frustration while at other times toward exhilaration and hours of pleasure. If media texts offer children moral certainty, this valuable characteristic of their play is weakened. Years later when political uncertainty confronts these children, now grown to adult citizen status, a rush toward certainty could mean death and destruction for thousands.

It strikes me that the largest part of the war toys' problem falls within the purview of adults—the adults who make, market, sell, and buy war toys but, even more so, the adults who do not make time to play with war toys and their children. Children need to learn new ways to play with toys so that they are not dependent on the sort of simplistic closed narratives Ebert suggests they need and with which the marketplace is all too happy to provide them. Whether traditional war narratives or tales of beauties and beasts, moral certainty makes for a bland ethical diet and undernourishes children's critical thinking skills. Children are wonderfully creative and can create the stories they need when given the necessary resources.

If adults allowed children age-appropriate leeway to experiment and play in a controlled environment rather than tried to protect and shelter them from topics deemed upsetting or too mature, then children could learn for themselves what place war and militarism should hold in their lives. I am convinced that children are intelligent enough and that war is horrible enough that most would make decisions with which the adults in their lives would be comfortable. The necessary resource this scenario depends on is that children will have access to an adult who is informed and thoughtful regarding the issues of war, violence, and militarism. Given the far-reaching effects of promilitarist ideology, I feel certain that my readers will acknowledge that all children do not have access to such an adult, in which case one of the following scenarios is probable: the children live in homes where war toys are not an issue and are, therefore, able to go about their play lives as they wish, which, in the majority of cases, means reproducing the narratives and morals of dominant ideology; the children become influenced by peers who do play with an informed adult and their play lives are wholly or in part transformed; and the children live with parents who prohibit war toys and are unable to learn play styles that could help them cope with and negotiate the realities of war and militarism. The last are the parents who can most productively be reached. They are concerned about their children's toys and play, and they are already sensitive to the unhealthy aspects of war entertainment.

The most egregious error parents who prohibit war toys are guilty of is the limited frame of reference they rely on when defining war toys.[27] War toys most commonly are comprised of objects that represent the means to conduct armed violence. This understanding of war toys grows out of a definition of war as armed combat between two or more parties in conflict. While these definitions seem rational enough, they are too limited for the twenty-first century. War in the current millennium includes economic sanctions that cripple and starve, national debt that forces a country into economic servitude, and environmental policies that damage a people's land and resources, making life in their traditional way and space no longer possible. Although humans may always have wars in which armed combatants face each other to do violence and destruction, more effective, wide-ranging, and socially acceptable ways to destroy a group's economic, political, and cultural lives are far more insidious. With such an understanding of war, a miniature bulldozer could be considered a war toy as could play money or a miniature bank. Any element of global capital could be, and I would argue should be, played with as if it were a war toy.

Playing with a scenario in which a bulldozer clears rainforest lands in order to make grazing land for beef to be packaged in a Happy Meal opens children's imaginations to the connections within which the world operates. Happily, bulldozers can also knock down multinational corporations in order to make room for urban playgrounds, and they can also push sand from one place to another. My point is that any toy's location in the unjust system of global capital can be exposed and built into a narrative that children should be introduced to in age-appropriate measures. War toys are far from the only, or perhaps even the main, problem in the twenty-first century toy box. The most lethal problem is that we are raising another generation who very well may be unable to recognize or negotiate the complex narratives and relationships that globalization and militarism facilitate. When Gil Mars suggests that Globotech export the Commando Elite toys to "some generals in Central America," he is introducing the elements of cultural imperialism and militarism into children's play vocabulary so that it may one day appear in their political vocabulary. This is a learning moment in *Small Soldiers* that parents can help children repeat and apply to other toys and play narratives.

Infiltrating Camp Commando

The unbelievably futuristic aspects of Globotech's Commando Elite toy soldiers seem less futuristic and unbelievable once we begin to study today's toy industry. The new smarter, more militarized toys of today demand more intelligent modes of interaction. Military and children's cultures share a performative and interpretative mode that would meet this demand: camp. According to Kerry Mallan and Roderick McGillis,

> Camp aesthetics disrupt or invert many Modernists' aesthetic attributes, such as beauty, value, and taste by inviting a different kind of apprehension and consumption. . . . A camp aesthetic delights in impertinence. It likes to challenge rather than satisfy. Its satisfactions derive from a sort of *puissance* of acceptance.[28]

In his influential study of gay and lesbian military personnel in World War II, Allan Berube described not only the lives of gay and lesbian people but also the performances of queer gender and sexuality that took place in many GI shows. Cross-dressing and playing with gender were vivid aspects of military life during this war and can offer a slantwise view of militarized gender today. The ability to disrupt militarized masculinity's dominating control over strength, courage, and patriotism lends camp a certain interpretive and cultural power. By exposing hypermasculinity as a performance, camp attests to the fact that there is nothing natural or normal about the alignment of militarized masculinity with the qualities it professes to control. With humor and impertinence, camp drives a wedge between militarism and hypermasculinity. This splitting is core to camp's influence as a cultural force.

Small Soldiers invites a camp reading in a number of ways: with its absurd exaggeration of military masculinity in the personas of the Commando Elite; its winking nods to Hollywood war films by referencing *To Hell and Back, Patton, Apocalypse Now,* and *The Dirty Dozen;* its gruesomely playful weaponization of the mundane in the combat scenes; and with its reference to the campy mockumentary *Spinal Tap,* by using the cast members from this film as the voice actors for the Gorgonites.[29] Added to this list of elements, Alan Abernathy's point of view with regard to the narrative invites viewers to entertain the notion that combat is not a performance of courage or patriotism, but a mindless destructive performance set into play by anonymous forces far behind the scenes who stand to profit from the show. Alan's viewpoint offers a third position for the children who may identify with him—neither perpetrator nor noble victim but witness to destruction and collateral damage survivor. The commodity universe of *Small Soldiers* is a play engine that invites the awareness that war is the cruel performance of programmed puppets whose consequences are real and hurtful. The film text forges an awareness of knowledge that U.S. military corporations manufacture and export toys that keep the dangerous game of war and the arms race in motion around the world. Alan, a not-so-tough kid who in the end gets the girl, or at least one of her kisses, offers the standpoint from which such an awareness can be pieced together and later applied to real-world scenarios.

Alan does not disrupt the programming of the Gorgonites, who set out like noble savages turned colonizers on a voyage to find their homeland, Gorgon. Whether in their quest they will displace metaphorical toy equivalents of Palestinians or will model a life of allegiance to a truly fictional mother country, perhaps the ideal example of Benedict Anderson's notion of an imagined community, is up to the collectors who buy and put into use all the *Small Soldiers* play sets and toys. What the movie has done is left a narrative base that is wide open and ambiguous in which children can explore and experiment. More experienced players can enter this world of play and introduce those with less experience to some of the issues contemporary global citizens must negotiate without lecturing on imperialism, militarism, or current events. Over time, play within this narrative will generate questions and connections that can be answered and developed as situations warrant without dampening fun or compassionate awareness.

A Sum Greater than Its Parts

Media conglomeration makes product releases like those that accompanied *Small Soldiers* more efficient and effective. When birthed within the subsidiaries of parent corporations such as Viacom, characters can go from product development to full market saturation in a media instant. Such was the case with Jimmy Neutron, whose television presence on Nickelodeon coincided with the Paramount Pictures movie release and the typical slew of electronic, personal, household, and toy products. Conglomeration intensifies the synergy around such product releases. Television episodes link immediately to computer game scenarios, and the character's key action pose in both is soon printed on a T-shirt. Media critics are now adept at interpreting the industry status quo that is conglomeration and the force that is synergy. Cultural critics need to become as adept with a vocabulary and critique for the convergence and synergy formations across industry, media, and government realms that have infiltrated militarist entertainment.

Technological innovation takes place more rapidly than language can account for. While the Cold War appellation "Military Industrial Complex" may have adequately described the convergence of management, research, and development that helped to fuel the arms race, twenty-first-century hookups are more flexible and complex than this nostalgic label can suggest. James Der Derian has attempted to map the structures that have replaced the military industrial complex in his book *Virtuous Wars*. In addition to charting the transitory nodes of the military-industrial-media-entertainment network (MIME-NET), Der Derian attempts to "study up close the mimetic power that travels along the hyphens."[30] One such power is computer simulation games' ability to make memories. During his tenure as a University of Central Florida professor and also director of the Media Convergence Lab, Christopher Stapleton, claimed that the military "are actually the visionaries of experiential media."[31] Visual technology that is operational on the Xbox and Playstation is also used by the army, with Pandemic Studio's *Full Spectrum Warrior* serving as the most current example.[32] This real-time tactics war game serves as entertainment in the private sector and command-tactical training in the military sector. Michael Macedonia, a technology officer for the army's Simulation, Training, and Instrumentation Command, understands the usefulness of current game technology this way:

> Essentially entertainment and games, that is, entertainment and training have an intersection: it's about making memories. It's fascinating now what we're learning about the human brain. . . . [A] lot of what we're trying to do in training is creating memories. Memories that last forever.[33]

Memories made while playing a military action game for fun could one day serve a young person in active duty with the military, their brains having been wired not only for survival but for command, evasion, and attack. As our toys become more intelligent, and therefore more like us, we become more like our toys, programmed to execute memories stored since youthful play sessions enjoyed long ago on the home front. William Hamilton's 2003 *New York Times* article titled "Toymakers Study Troops, and Vice Versa" introduced readers to the flow of research and development information that takes place between the toy industry and the military industry.

Hamilton cited such cross-pollination examples as Hasbro's Super Soaker on which the army based its quick-loading assault weapons design, remote control planes that have inspired reconnaissance drones, and inexpensive cell phones that became the basis for walkie-talkies with video capability. In products that could have been developed by Globotech, the control mechanisms for unmanned robotic vehicles, some of which are used in Iraq to deactivate improvised explosive devices (IEDs), have been based on video game controllers. Lance Winslow's military strategy report "Unmanned Vehicle Robotic Warfare" imagines:

> Unmanned Aerial Vehicles . . . making human fighter pilots obsolete. Might be better to use a UAV and to that have it flown by a 16-year old video game player, which will in turn be used to program the next generation of Artificial Intelligent robotic tactical UAVs.[34]

Winslow's rapidly attainable fantasy calls up images of Ender Wiggin, the protagonist in Orson Scott Card's novel, *Ender's Game*, who unwittingly destroys an entire species and its home planet, believing he is only being tested on a computer simulation training program. For Ender and the generations of citizens raised on Nintendo-style war, technology gives the false impression that war can be surgically clean and takes place in far and distant lands; furthermore, technology exudes the seductive promise that it is safe to play with militarist violence on the home front. *Small Soldiers* is the only successful children's mainstream U.S. narrative that I know of to disrupt technology's lies; when technology that has been developed for the military enters the domestic sphere, whether in the form of a sophisticated computer chip in a toy with artificial intelligence or in a Super Soaker water gun, war and militarism gain in destructive power and civilians and citizens lose.

Conclusion

A policy of saying no to war toys has many drawbacks, one of which is the reality that enforcing any prohibition requires a force equal to or greater than the force of the desire for what is prohibited. I would imagine that most adults have witnessed children's performative tantrums once denied what they most want. Whether in a store, another child's play area, or in their own backyard, children's desire is a force with which to be reckoned. Sometimes desire can be rechanneled and children's attention fixed on an acceptable alternative toy or form of play. However, when this does not work, when children's fixation on a war toy or war play is parallel to the adult world's fixation with war, then the force of prohibition must overcome the children's desire. This is one of the most difficult terrains of child rearing, and it is territory that in any militarized society is in danger of being dominated by a logic of power informed by militarist values and strategies: punishment or the threat of punishment is the only way to resolve a conflict and the world is divided up into two distinct categories, good and bad, where the bad is inherently evil and must be suppressed by the good.

Rather than try to thwart children's fascination with war play and war toys, it makes more sense to move with the force of their desire and learn to challenge, redirect, subvert, and accept the products of their imaginations in context and over time. Play is a fertile realm for understanding, not only history and politics, but the emotions of being human in a complex, highly militarized world. Children must learn equally complex strategies of social negotiation and emotional coping on the road to becoming engaged global citizens. In addition to making wiser, more connected, and more aware citizens, compassionate and engaged war play has a further advantage in that it can become a realm in which the work of demilitarization can begin. The process of demilitarizing any one or all nations will be enormous and multifaceted. Workers will need to be retrained, desire will need to be redirected, factories will need to be refitted, and identities will have to be refigured. Toys and the world of play can become one realm where this massive project could begin to be imagined. Play is a creative place cordoned off from many of the limiting forces of reality. The demobilization and refunctioning of MIME-NET is too large a project to imagine, let alone to undertake. However, demobilizing the forces of militarism in the world of toys seems far more possible. If we cannot begin here, then where?

Notes

1. John Fiske, *Understanding Popular Culture*. Boston, 1989, p. 14.
2. Darlene Hammell and Joanna Santa Barbara, "War Toys/PGS Briefing Paper," in *Physicians for Global Survival*. http://www.pgs.ca/pgs.php/prevention/11/. Accessed 30 October 2007. PGS is the Canadian chapter of International Physicians for the Prevention of Nuclear War.
3. For an example of militarist-styled children's bedding: BabyUniverse.com's California Kids Flying Tigers collection with its "patchwork of khaki planes, green planes, solid chambray, camouflage and solid cream [that] make up this WWII inspired bedding collection. This set features lots of high flying action for the little soldier in your life" (*BabyUniverse*. http://www.babyuniverse.com/ kit/baby/1148/FlyingTigers.html. Accessed 30 October 2007).

4. "Pacifist Mamas Ban Toy Soldier Wars They Rule That Wooden Guns Are Menace to World Peace," *Washington Post*. 12 December 1933, p. 15.

5. John Michlig, *GI Joe: The Complete Story of America's Favorite Man of Action*. San Francisco, 1998, p. 155.

6. The once trusted name in creative nonviolent toys, now even Lego produces play sets that promote fantasy fighting such as the Star Wars, Galidor, and Exo-Force series.

7. One early example of a crusader against war toys was Constance Wilde, wife of Oscar Wilde, who addressed the Women's Committee of the International Arbitration and Peace Association in 1888. Some years later, activism among toy pacifists was so popular that the famed short-story writer Saki wrote a piece entitled "The Toys of Peace," which satirized the trend (Ed Halter, *From Sun Tzu to Xbox: War and Video Game*. New York, 2006, pp. 50, 54).

8. Daphne White, "From War Chests to Toy Chests: How to Change your Child's Worldview for the Better, One Toy at a Time," in *Mothering*. http://www. mothering.com/articles/growing_child/consumerism/toy_chests.html. Accessed 5 May 2008.

9. Patricia Cohen, "Child's Play Has Become Anything but Simple," *New York Times*. http://www.nytimes.com/2007/08/14/books/14play.html. Accessed 5 May 2008.

10. Elizabeth Gardner, "Understanding The Net's Toughest Customer: Why It's Worth Targeting Those Elusive 13-and-Unders," *Internet World Magazine*. http://www.iw.com/magazine.php?inc=020100/2.01coverstory.html. Accessed 5 May 2008.

11. "Movie Cross Promotions," *Discount Store News*. http://findarticles.com/p/articles/mi_m3092/is_16_37/ai_50267681. Accessed 5 May 2008; Suna Chang, "The Toys of Summer," *EW.com*. http://www.ew.com/ew/article/0,,284896,00. html. Accessed 5 May 2008.

12. Dale Buss, "A Product-Placement Hall of Fame," *Business Week*. http://www. businessweek.com/1998/25/b3583062.htm. Accessed 5 May 2008.

13. Julianne Hill, "The PG-13 Trap," *Promo*. http://promomagazine.com/entertain-mentmarketing/marketing_pg_trap/. Accessed 5 May 2008.

14. Richard Morgan, "Size Mattered for Product Tie-Ins," *Variety*. http://www. kobinenterprises.com/webpages/presspages/rayban.html. Accessed 5 May 2008.

15. Jonathan Rosenbaum, *Movie Wars: How Hollywood and the Media Limit What Movies We Can See*. Chicago, 2002, p. 67; Chang, "Toys of Summer."

16. Frederic Jameson, *Postmodernism, or, The Cultural Logic of Late Capitalis*. New York, 1991; Jean Baudrillard, *Simulation*. New York, 1983; Jane Kenway and Elizabeth Bullen, *Consuming Children: Education-Entertainment-Advertising*. Philadelphia, 2001; Kim Humphery, *Shelf Life: Supermarkets and the Changing Cultures of Consumption*. Cambridge, 1998.

17. Gisela Wegener-Spöhring, "War Toys in the World of Fourth Graders: 1985 and 2002," in Jeffrey Goldstein, David Buckingham and Giles Brougère, eds., *Toys, Games, and Media*. Mahwah, New Jersey, 2004, p. 31.

18. Ibid.
19. Roger Ebert, "Small Soldiers," *RogerEbert.com*. http://rogerebert.suntimes.com/ apps/pbcs.dll/article?AID=/19980710/REVIEWS/807100306/1023. Accessed 5 May 2008.
20. Research conducted by Children's Research Unit report, *Youthsight*, suggests that even the youngest viewers will understand that the film's opening is a commercial (in Kenway and Bullen, p. 110): "Children from as young as three years old can recognize the persuasive intent of advertising. While at this early age, they can verbalise the role of advertising as 'they're trying to get me to buy it,' by the age of five or six this has developed into 'they are trying to sell me something.' By seven, most children are capable of understanding exactly what advertisers are trying to achieve and, by then, children have become adept critics and prove a hard—even cynical—audience to influence."
21. Ebert, *Small Soldiers*.
22. Scott Rosenberg, "Toy Gory," *Salon.com*. http://www.salon.com/ent/movies/ reviews/1998/07/10review.html. Accessed 5 May 2008.
23. Ed Halter, "War Games: New Media Finds Its Place in the New World Order," *Village Voice*. http://www.villagevoice.com/news/0246,halter,39834,1.html. Accessed 5 May 2008; Halter, *From Sun Tzu*.
24. Krista Foss, "All I Want for Christmas Is a Bombed-Out Dollhouse," *Globe & Mail/Canada*. http://www.commondreams.org/views02/1123-01.htm. Accessed 5 May 2008.
25. Israel Canlapan, "*Small Soldiers*, Bi Ho-Hum," *Christian Answers.net*. http://www.christiananswers.net/spotlight/movies/pre2000/i-smallsoldiers.html. Accessed 5 May 2008.
26. Wegener-Spöhring, "War Toys in the World of Fourth Graders," p. 25.
27. The brightly colored handmade puppet sewn in war-torn Sri Lanka is as much a war toy as the frighteningly realistic and powerful paintball gun; one's connection to violence and injustice is more hidden than the other's, but both are caught in the web that is militarism. I find just as distressing that children and adults cannot make the connections between seemingly innocuous products and global violence as the fact that children play gleefully with war toys as they are traditionally understood.
28. Kerry Mallan and Roderick McGillis, "Between a Frock and a Hard Place: Camp Aesthetics and Children's Culture," *Canadian Review of American Studies/ Revue canadienne d'etudes americaines*. 35, 1, 2005, p. 3.
29. Jesse Hibbs, dir., *To Hell and Back*. Universal, 1955; Franklin J. Schaffner, dir., *Patton*. Twentieth Century Fox, 1970; Francis Ford Coppola, dir., *Apocalypse Now*. Zoetrope Studios, 1979; Robert Aldrich, dir., *The Dirty Dozen*. MGM, 1967; Rob Reiner, dir., *This Is Spinal Tap*. Spinal Tap Productions, 1984.
30. James Der Derian, *Virtuous War: Mapping the Military-Industrial-Media-Entertainment Network*. Boulder, 2001, p. xx.
31. Halter, *From Sun Tzu*, p. 194.

32. Clive Thompson, "The Making of an X Box Warrior," *New York Times.* http://query.nytimes.com/gst/fullpage.html?res=9C02EEDD133FF931A1575BC0A9629C8B63. Accessed 5 May 2008.

33. Halter, *From Sun Tzu,* p. 198.

34. Lance Winslow, "Unmanned Vehicle Robotic Warfare," *The Online Think Tank.* http://www.worldthinktank.net/pdfs/unmannedvehiclerobotic.pdf. Accessed 5 May 2008.

Notes on Contributors

Maartje Abbenhuis is a senior lecturer in modern European history at the University of Auckland. She received her PhD in history from the University of Canterbury, New Zealand, in 2001. Her research interests are the history of European neutrality from the Napoleonic era to the Second World War and historical investigations of borderland theory. Her first book, *The Art of Staying Neutral: The Netherlands in the First World War*, was published by Amsterdam University Press in 2006.

Penelope Adams Moon presently holds the chair in the history department at Bethel College in North Newton, Kansas. She received her PhD in history from Arizona State University in 2001. Her publications include "'Peace on Earth—Peace in Vietnam': The Catholic Peace Fellowship and Antiwar Witness, 1964–1976," *Journal of Social History,* 36, Summer 2003, pp. 1033–1057; and "Loyal Sons and Daughters of God?: American Catholics Debate Catholic Antiwar Protest," *Peace & Change,* January 2008, pp. 1–30.

Irene Andersson is a lecturer in history at the University of Malmö. She has studied Swedish women's networks and collective peace actions during the First World War and in the 1930s. Her PhD thesis (in Swedish, with an English summary) is titled "Women Against War: Actions and Networks for Peace 1914–1940," Lund, 2001. Among her work is an article on Fredrika Bremer's invitation to form a Peace Alliance in 1854 and how the Swedish Women's Movement has used it during the last century and an article on how the Army museum in Stockholm explains violence in their permanent exhibition. Forthcoming is a book on gender and surveillance of peace activists by the military's secret service, 1930–1970, with the working title *Idealists or Cryptokommunists: Gender Perspectives on Surveillance of Peace-Workers in Sweden during 1930–1970.*

Sara Buttsworth received her PhD from the University of Western Australia in history and women's studies in 2004. Sara has been interested in issues of gender and warfare since her contribution of a general chapter on women and the Second World War in Western Australia to a collection titled *On the Homefront* that was published in 1996. More recently she has published "'Bite Me': *Buffy* and the

Penetration of the Gendered Warrior Hero," *Continuum,* 16, 2, 2002, pp. 185–199; and "Who's Afraid of Jessica Lynch? Or One Girl in All the World?: Gender, Heroism and the Iraq War," *Australasian Journal of American Studies,* 24, 2, December 2005, pp. 42–62. Sara's first book, *Body Count: Gender and Soldier Identity in Australia and the United States,* was published by VDM Verlag Dr. Müller in 2007. Sara has been teaching in the Department of History at the University of Auckland since 2004, mainly in the Tertiary Foundations Certificate Programme.

Gabrielle A. Fortune is an honorary research associate in the Department of History at the University of Auckland, New Zealand. She received her degrees from the University of Auckland, most recently her PhD in 2005. Her research interests include war brides (the subject of her PhD dissertation), war veterans, and the appropriation of war images in popular culture. Together with her colleague, Mara Bebich, Gabrielle has just published a short history of returned New Zealand servicemen, titled *Citizenship & Remembrance: A History of the 24NZ Infantry Battalion Association.* Auckland, 2008.

Karen J. Hall publishes on militarist material culture produced predominantly for a U.S. audience. She is currently a humanities post-doctoral faculty fellow in the English Department in the College of Arts and Sciences at Syracuse University. Karen's most recent publications reflect her passion for this subject area: "Photos for Access: War Pornography and US Practices of Power," in Nico Carpenter, ed., *Culture, Trauma & Conflict: Cultural studies Perspectives on Contemporary War,* Cambridge, 2007; "Consuming Witness: Combat Entertainment and the Training of Citizens," in Frances Guerin and Roger Hallas, eds., *The Image and the Witness,* London, 2007; "Shooters to the Left of Us, Shooters to the Right: First Person Arcade Shooter Video Games and the Violence Debate," *Reconstruction: Studies in Contemporary Culture,* 5, supplement, December 2005; "A Soldier's Body: G.I. Joe, Hasbro's Great American Hero and the Symptoms of Empire," *Journal of Popular Culture,* 38,1, 2004.

Suellen Murray is a senior research fellow at the Centre for Applied Social Research at RMIT University in Melbourne, Australia. Her main research interests are concerned with historical analyses of domestic violence and social policy about which she has published widely, including her book *More than Refuge: Changing Responses to Domestic Violence,* University of Western Australian Press, 2002. Her involvement in political action around violence against women and later academic work in this area emerged from her participation in the women's peace movement in the 1980s, a theme to which she returns in her chapter in this book.

Mark A. Potter graduated with a PhD from the University of Melbourne, Australia, in 2005. His thesis, "A Good Soldier, a Good Shot, and a Good Fellow: The Seventy-first New York, Martial Manhood and the Shadows of Civil War, 1850–1898," suggested that martial enthusiasm in late-nineteenth-century America may have had a life outside of social upheaval, the crisis of masculinity, and sectional reconciliation. Mark is currently working in research management while applying for postdoctoral funding. His publications include "A Pomp that Cannot be Distinguished from Real War: Gilded Age Martial Display," in Katherine Ellinghaus, David Goodman and Glenn Moore, eds, *Unsettling America: Crisis and Belonging in United States History,* Melbourne, 2004, pp. 91–102.

David M. Rosen is professor of anthropology and law at Fairleigh Dickinson University in Madison, New Jersey. He received his PhD from the University of Illinois, Urbana-Champaign, and his JD from Pace University School of Law. He has carried out field research in Sierra Leone, Kenya, Israel, and Palestine. His prime interests are in the relationship between law and culture and in the anthropology of children and childhood. He is the author of *Armies of the Young: Child Soldiers in War and Terrorism,* New Brunswick, NJ, 2005. He is currently carrying out research on the war crimes trials in Sierra Leone.

Ismee Tames is a researcher at the Netherlands Institute for War Documentation in Amsterdam. Her PhD thesis "War on Our Minds": War, Identity and Neutrality in Dutch Public Debate, 1914–1918 was published in May 2006 (in Dutch— article in English forthcoming). She is currently working on a new project on "Children of Collaborators in the Netherlands, 1945–1960."

Index